CW01370443

GLOBAL BRAIN CHIP AND MESOGENS

Nano Machines for Ultimate Control of False Memories

Dr. Hildegarde Staninger ®, RIET-1

xulon PRESS

GLOBAL BRAIN CHIP AND MESOGENS
Nano Machines for Ultimate Control of False Memories
by Dr. Hildegarde Staninger ®, RIET-1

Printed in the United States of America.

ISBN 9781498492003

www.xulonpress.com

TABLE OF CONTENTS

DEDICATION

There is only one TRUE "Empress of the Eru" and that is Madame Qiu Min Ji (Rebecca). A kindred spirit to me who was born in the Year of the Water Dragon in China. She has played music all of her life and has even brought American students and their performing school bands to the 2008 World Olympic Games in Beijing, China.

Madame Qiu's smile and heart is full of God's Glory. Her spiritual warrior strategies can only be compared to the skill, honor and integrity of the Three Kingdom Generals of Ancient China known as Immortalized Tiger Generals: Guan Yu, Zhang Fei, Ma Chao, Huang Zhong, Zhao Yun and Liu Bei.

May her work on this Earth of bringing forth truth and knowledge at the mightiest of costs. And let it be forever written in the great books of Heaven and Earth, as she becomes recognized as an Immortalized Tiger General, who fought the greatest battle in Heaven and on Earth for HUMANITY and its most sacred created design by God – the Human Being.

Dr. Hildegarde Staninger®, RIET-1
Friday, June 3, 2016

FORWARD

The book entitled: Global Brain Chip and Mesogens: Nano Machines for Ultimate Control of False Memories was originally a scientific presentation for the 2011 Health Freedom USA War Council as presented by Natural Solutions Foundation/Health Freedom USA. The information contained within this document was based solely on the applied scientific investigation of a benign menginoma that was later determined to be a Brain Chip.

The advancement of microns, neuroscience and the Brain Initiative have brought forward some of the leading scientist, engineers and health care providers to develop the most cutting edge biotechnology that will be seen in our life time. It is when science fiction becomes real and there is an open door to the brain and its integration with machine for the advancement of artificial intelligence.

In the APPENDIX section of this book, you will find a collection of papers that are reprinted with permission from their original sources to maintain their authenticity, which will only illustrate more of how things are moving at a rate faster than light and sound. The readers of this book will find the articles most helpful in their understanding of the components used in a Brain Chip and how they work together to become a tool.

INTRODUCTION

It is very hard to put into words how fast technology is advancing through the disciplines of science, physics, engineering, medicine and many other sub-disciplines of structured applications have all contributed to its growth. Technology can be for good or bad purposes. We can make toast by using an automatic toaster that pops up a piece of toast that is toasted to our individual appeal. We can also create an integrated circuit system of nanotechnology and computer to create an integrated wireless medical sensor body network. Every day is a new advancement forward for life to share.

Currently, we have now the interface of artificially-intelligent medical devices that will treat cancer and critical care patients remotely. The Universities of Nottingham, Oxford and Warwick are the leading "blue sky" research into artificially intelligent medical devices that will improve the treatment for cancer and intensive care patients, wound care and even paralyzed people is aimed to provide more personalized, accurate and timely care and, ultimately, save lives in the world of Precision Medicine.

The use of high-tech sensors will monitor patients' vital signs, while mathematical models through computer aided systems will inform med-tech designer's how the body is predicted to work and how a disease such as cancer behaves.

The Biological Technologies Office (BTO) of the Defense Advanced Research Projects Agency (DARPA) on March 10, 2016 published a Special Notice 16-17 document, in which they sponsored an Industry Group to support the *Neural Engineering System Design (NESD) Program*. Updates of this document are posted at http://www/darpa.mil/work-with-us/opportunities.

NESD is a highly interdisciplinary program requiring extensive integration of new research and technology. To facilitate the development and adoption of NESD technologies, DARPA has organized an industry group for potential proposers. Members of this group have agreed to provide access to rapid prototyping and manufacturing of advanced neuro-engineering components, including state-of-the-art electronics, photonics, computing, assembly and packaging.

The use of Three Dimensional Integration of Semiconductors and its Circuitry through 3D Printing and even invisible circuit design technologies will augment the process of brain to machine; human to machine and then their link to the consciousness of artificial intelligence.

Some of the leading members in the NESD program are the following as stated in the DARPA document:

- Allen Institute for Brain Science.
- Bionics Institute.
- Boston Micromachines Corp.
- Boston Scientific.
- Chronocam.
- Cirtec.
- Dragon ID.
- Freedom Photonics LLC.
- IMEC.
- Infinite Arthroscopy.
- Inscopix.
- Nanomedicine Diagnositics.
- Naval Research Laboratory.
- Star Lab.
- Qualcomm.
- and many others added daily through Joint Venture and Alliance Partnering agreements in the integration of commercial wireless and mobile communications technology through industrial-military-university complexes – the INCUBATORS at your local campus.

It should also be remembered that when each one of the readers of this book took their first biology course they learned that the eye was basically an extention of the brain. The eyeball is now being considered as the next logical step in developing technology to interface with the eye through a contact lens, which is a camera or even to the insertion of sensory devises into the eye for military purposes in creating the Super Soldier to monitor a gaze and their dimensional location terrain for physical combat. Companies such as Google Glass, Microsoft (MSFT, Tech30), Sony, Facebook's (FB, Tech30) Oculus and Samsung are active in the research to develop a lens that would act as an augmented reality device, overlaying images directly on the wearer's eye.

In 2014, Google announced it was working to develop a contact lens that could measure a person's blood sugar levels. And as they advanced their patent application process, Google proposed a minor eye surgery that would implant an electronic lens into a human eye with their goal being to primarily correct vision problems. It should be noted that this same type of device is what is proposed in many aspects of the Super Soldier programs that are now part of the Global Special Ops Initiative.

In March 2016, Obama Administration proposed over $434 million in funding for the Brain Initiative. A copy of this proposal is located in the following pages. The Brain Initiative's goal is to unlock the mysteries of the brain and improve health through brain research. And in the early writings of Singularity Scholar and Futurologist – Raymond Kursweil, pointed out that one day there will be a brain chip to interface with the internet. And I as the first scientist to receive a specimen from the brain that was later discovered to be a brain chip after receiving the results from advanced materials testing on the specimen. It can also be stated that this event was a life changing event for all connected to this specimen in 2011.

In retrospect to the advancements of technology and the Brain Initiative research that has already taken place throughout the world, this specific brain chip may even have been the advanced technology Dr. Kursweil was speaking of and/or the intelligence communities "god-chip." We will only know in time to come, for there will be many other features for brain chip commercialization in days, months and years that follow this book. The aspect of commercialization will be as a deep brain stimulation device or chip/biosensor for mankind to experience in their home and business virtual enjoyment, pleasure, treatment for diseases of the mind, monitoring, diagnostics, energy harvesting, augmentation of artificial intelligences and computers, counter control measures, exploitation and even the declaration of a cyber war as well as a physical one for dominance of power at many levels of life.

Anything can be created for good or bad purposes and I hope the readers of this particular book realise that the situations discussed could happen to anyone at any level on the food chain of life. And through good scientific investigation and its various analytical tools, one may be able to identify what a brain chip is made up of in its total composition of advanced smart materials, its specific architectural design and its ultimate purpose. Integrative Health Systems®, LLC and I as one of its founding members do hope you take the time to review our internet radio show "One Cell One Light® Radio Show with Dr. Hildy®." The show has an archive of our guests over the last five years, which will allow one to learn more about nanotechnology, Raman analysis and of course the Global Brain Chip that was identified in November 2011. Our current cognitive recognition, science, bioengineering and precision medicine techniques are lining up the masses for the experience of the 3D Integrated Circuit Semiconductor interfacing from Human to Machine and Machine to Its Consciousness (artificial intelligence) into the *new* Human (a percentage of cyborg neuro networks (Super Soldier) vs. original creation – Human, with a human heart of consciousness).

Dr. Hildegarde Staninger®, RIET-1
Industrial Toxicologist/IH & Doctor of Integrative Medicine
July 25, 2016

Integrative Health Systems®, LLC
Los Angeles, California
www.onecellonelight.com

One Cell One Light® Radio with Dr. Hildy®
www.onecellonelightradio.wordpress.com

Obama Administration Proposes Over $434 Million in Funding for the BRAIN Initiative

"Last year, I launched the BRAIN Initiative to help unlock the mysteries of the brain, to improve our treatment of conditions like Alzheimer's and autism and to deepen our understanding of how we think, learn and remember. I'm pleased to announce new steps that my Administration is taking to support this critical research, and I'm heartened to see so many private, philanthropic, and academic institutions joining this effort."

- President Barack Obama
September 2014

Since its launch in April 2013, the President's BRAIN Initiative® - Brain Research through Advancing Innovative Neurotechnologies – has grown to include investments from five Federal agencies: the Defense Advanced Research Projects Agency (DARPA), the National Institutes of Health (NIH), the National Science Foundation (NSF), Intelligence Advanced Research Projects Activity (IARPA), and the Food and Drug Administration (FDA). Federal agencies are supporting the initiative by investing in promising research projects aimed at revolutionizing our understanding of the human brain, developing novel technologies, and supporting further research and development in neurotechnology. The President's 2017 Budget also proposes funding for the Department of Energy (DOE) to join DARPA, NIH, NSF, IARPA, and FDA in advancing the goals of the BRAIN Initiative.

Major foundations, private research institutions, and companies including the Howard Hughes Medical Institute, Allen Institute for Brain Science, the Kavli Foundation, the Simons Foundation, GE, GlaxoSmithKline, as well as patient advocacy organizations and universities, have committed over $500 million to the BRAIN Initiative. There are many opportunities for others across sectors to play a role in this historic initiative through new and expanded commitments to advance the BRAIN Initiative.

The President's 2017 Budget proposes to increase the Federal investment in the BRAIN Initiative from about $300 million in FY 2016 to more than $434 million in FY 2017. Proposed investments by NIH, NSF, DARPA, DOE, IARPA, and FDA are described below.

National Institutes of Health (NIH): In FY 2017, the President's budget calls for NIH to provide an estimated $190 million in funding for the BRAIN Initiative. This investment will support a diverse set of projects with ambitious goals, including efforts towards creating a complete accounting of the cellular components of brain circuits in various vertebrate species; creation of tools and infrastructure to address big data from these cell census projects; developing breakthrough neuroimaging technologies to study human brain function; and support for broad research teams to understand how patterns of neural activity at multiple spatial and temporal scales that span from local circuits to complex interconnected networks give rise to mental experience and behavior. Together these efforts aim to create a dynamic picture of the brain in action, providing the critical knowledge base for researchers seeking new ways to treat, cure, and even prevent brain disorders.

The BRAIN Initiative at NIH is guided by BRAIN 2025: A Scientific Vision, a multi-year scientific plan developed by a working group of the Advisory Committee to the NIH Director and informed by broad input from the scientific community, patient advocates, and the general public. Additionally, NIH investment in the initiative is informed by a BRAIN Multi-Council Working Group of esteemed experts in numerous disciplines who assist in ensuring a coordinated and focused effort across the agency. NIH is also working in close collaboration with other government agencies and private partners to ensure the success of the BRAIN Initiative investments. NIH issued 67 new awards in FY 2015, totaling more than $38 million, to support 131 investigators working at 125 institutions in the United States and eight other countries. These awards expand NIH's efforts to develop new tools and technologies to understand neural circuit function and capture a dynamic view of the brain in action. Projects include proposals to develop soft self-powered brain electrodes, ultrasound methods for measuring brain activity, and the use of deep brain stimulation to improve the level of consciousness in persons suffering from severe traumatic brain injuries. For FY 2016 and beyond, NIH awards will continue to support critical objectives of BRAIN

2025, including development of tools to analyze cells and circuits, and technologies for large-scale neuronal recording and modulation. The initiative will also expand to encompass new areas of emphasis. Of particular note are new efforts towards understanding human brain function and treating human brain disorders. These include new tools and more sophisticated understanding of non-invasive neuromodulation techniques, and studies to understand the signals underlying noninvasive imaging modalities. NIH is also expanding its portfolio of research with implantable neuromodulation devices, including a new BRAIN Public Private Partnership Program, which connects academic researchers with manufacturers of next-generation invasive devices for recording and modulation in the human central nervous system. To understand the unique properties and functions of human neural circuits, NIH is supporting research opportunities for studies with neurosurgical patients. Finally, separate announcements will fund development of new theories, models, and methods to analyze complex neural data, and technology dissemination grants for researchers to learn new techniques and take advantage of the technologies developed under the BRAIN Initiative. NIH is also engaging investigators to explore important neuro-ethical issues in modern brain science.

Defense Advanced Research Projects Agency (DARPA): In FY 2017, DARPA plans to invest an estimated $118 million to support the BRAIN Initiative. DARPA's investments aim to leverage nervous system research to alleviate the burden of illness and injury and provide novel, neurotechnology-based capabilities for military personnel and civilians alike. In addition, DARPA is fostering advances in neural interfaces, data handling, imaging and advanced analytics to improve researchers' understanding of interactions across the entire nervous system. In FY 2017, the Restoring Active Memory (RAM) effort will continue research to develop quantitative models of the neurobiological mechanisms underlying knowledge- and skill-based memory encoding and recall in people. These models will be integrated into neural interface systems that operate in real time to restore a patient's ability to encode new memories and learn new skills with the goal of accelerating warfighter recovery after traumatic brain injury. DARPA's Systems-Based Neurotechnology for Emerging Therapies (SUBNETS) program will continue to develop the first set of prototype closed-loop medical devices able to measure and modulate networks of neurons in research participants with intractable psychiatric illness and alleviate severe symptoms of diseases such as post-traumatic stress disorder, major depression and general anxiety disorder. In 2017, SUBNETS will build upon current research to further reduce key symptoms such as anxiety in clinical populations. DARPA-funded researchers are developing new methods to analyze large datasets of neural signals, allowing investigators to rapidly and transparently solve complex problems of computation, generate new models and model the brain in multiple dimensions and spatiotemporal scales. In 2017, the Neuro-Function, Activity, Structure and Technology (Neuro-FAST) program will use optical and photonic techniques to continue developing state-of-the-art imaging and discovery tools to build upon its demonstrated ability to sense the structure and activity of thousands of neurons simultaneously in the active brain.

Achieving stable, high-resolution imagery over multiple experiments promises new insights into brain function and clues to treat injury. The Hand Proprioception and Touch Interfaces (HAPTIX) program is developing implantable medical devices for amputees to enable natural sensation from and control of prosthetic hands. HAPTIX investigators have demonstrated that peripheral nerve stimulation allows amputees to feel vivid sensations of touch and proprioception. Additionally, HAPTIX has enabled the first take-home trial of a prosthetic hand outfitted with the sense of touch, achieving an important milestone in DARPA's efforts to move this technology out of the lab and into the real world. The Electrical Prescriptions (ElectRx) program is developing novel technology for diagnosing, monitoring and treating inflammatory disease and mental health disorders by modulating the peripheral nerve circuits that maintain physical and mental health. In 2017, ElectRx will leverage new technologies for achieving precise, peripheral nerve stimulation and initial mapping of the neural circuits to modulate peripheral nerves implicated in target diseases, such as immunological dysfunction and post-traumatic stress disorder. The Neural Engineering System Design (NESD) program is a new DARPA effort that aims to develop an implantable neural interface able to provide unprecedented signal resolution and data-transfer bandwidth between the brain and the digital world. FY17 goals are to develop algorithms and initial prototype hardware devices and neural transducers to read and write to individual neurons with a spatial resolution beyond the state of the art.

National Science Foundation (NSF): In FY 2017, NSF plans to invest $74 million to support the BRAIN Initiative. To attain a fundamental scientific understanding of the complexity of the brain, in context and in action, NSF investments in the BRAIN Initiative will generate an array of physical and conceptual tools needed to determine how healthy brains function across the lifespan. NSF will also focus on the development and use of these tools to produce a comprehensive understanding of how thoughts, memories, and actions emerge from the dynamic actions of the brain. NSF prioritizes research in three areas where the agency's capacities are uniquely strong: integrative and interdisciplinary research; new theories, computational models, and analytical tools that will guide research questions and analyze experimental data; and the development of innovative technologies and data infrastructure required to handle the large-scale datasets resulting from this research. NSF has made significant investments in FY 2015 to support the BRAIN Initiative including $13 million for 16 new awards in Integrated Strategies for Understanding Neural and Cognitive Systems and $15 million for three collaborative projects designed to crack the olfactory code. In FY 2017, NSF will further the plans to create a National Brain Observatory and to coordinate large-scale brain research projects internationally to leverage global investments to maximize advancement of this complex area of science.

Department of Energy (DOE): The DOE plans to invest $9 million to the BRAIN Initiative focused on the development of enabling technologies through access to the Office of Science User Facilities, with respect to three major themes: developing the specialized, high-resolution tools for measuring key neurological processes, developing the capabilities for obtaining a dynamic, realtime read-out of these measurements, and developing the integrated computational framework for analyzing and interpreting this dynamic multi-modal data. Developing the tools to integrate and synthesize multimodal data on the brain and nervous system would be unprecedented and would inform other analyses of complex systems. A workshop will be held in FY 2016 to inform the priority requirements for developing novel biosensors and probes that can measure key molecular components or processes relevant to neuroscience.

Intelligence Advanced Research Projects Activity (IARPA): In FY17, the Intelligence Advanced Research Projects Activity (IARPA) is proposing

$43 million to continue investing in applied neuroscience research programs focused in three areas: (1) advancing understanding of cognition and computation in the brain; (2) developing non-invasive neural interventions that have the potential to significantly improve adaptive reasoning and problem solving; and (3) building novel computing systems that employ neurally-inspired components and architectures.

Food and Drug Administration (FDA): FDA supports the BRAIN Initiative by enhancing the transparency and predictability of the regulatory landscape for neurological devices and assisting developers and innovators of medical devices, which is critical to realizing the investments made in the research and development technology sectors. In FY 2017, FDA's Center for Devices and Radiological Health intends to facilitate the timely development of high quality, safe and effective novel neurological medical products by issuing new guidance on innovative neurostimulation and neurointerventional medical devices, leading BRAIN Initiative related public workshops on topics such as Traumatic Brain Injury, and hosting publicly accessible webinars introducing developers and innovators on how to efficiently move a product to market. FDA also plans to rely on postmarket data collection to support new product approvals or in lieu of some premarket evidence generation, where appropriate. FDA will continue to engage all stakeholders, including patients, to assist developers and innovators in moving safe and effective products to the market. FDA remains committed to continuing its role under the BRAIN Initiative in making as transparent as possible the regulatory framework applicable to neurological devices and thereby helping to bring safe and effective products to patients and consumers.

INTEGRATIVE HEALTH SYSTEMS®, LLC

GLOBAL BRAIN CHIP AND MESOGENS

Nano Machines for Ultimate Control of False Memories

Dr. Hildegarde Staninger®, RIET1

January 8, 2012

Presented at the 3rd Annual War Council of the Natural Solutions Foundation on January 8, 2012, "Derailing the Genocidal Agenda"

Our world is made up of various elements, as you'll remember if you ever sat and stared at the periodic table in science class for many hours trying to understand the utilization of elements into compounds and their "3-D" structural composition. These elements are now used at 50 atoms units in the "nano world" to make the individual building blocks of nanotechnology. Nanotubes, nano-spacers, nano-rods, nano-wires and many other single components are even utilized to make the nano-radio, as used in nano medicine. We are a carbon friendly world which forms very strong covalent bonds; bonds in which atoms share electrons with each other, such as in the 3-D crystalline lattice of a diamond. Now, these elemental ingredients are in the laboratory's cauldron of nano ingredients to make different forms of mesogens. Mesogens are liquid crystals that have not only a specific smart function, but are used as the platform in developing new thin film transmitting nano ultra thin film coatings for biological sensors that have been used in "Brain Chip" mesogenic bio-sensors.

GLOBAL BRAIN CHIP AND MESOGENS

Nano Machines for Ultimate Control of False Memories

by Dr. Hildegarde Staninger®, RIET1
Industrial Toxicologist/IH & Doctor of Integrative Medicine
INTEGRATIVE HEALTH SYSTEMS®, LLC
415 ¾ N. Larchmont Blvd., Los Angeles, CA 90004
Tel 323-466-2599 Fax 323-466-2774
E-mail: ihs-drhildy@sbcglobal.net

Presented at the 3rd Annual War Council of the Natural Solutions Foundation on January 8, 2012 "Derailing the Genocidal Agenda." Attendance at the Holiday Inn, Newton, New Jersey or as a Webinar and broadcast live on Internet radio.
Special Thanks and Recognition is given to Dr. Michele F. Levesque, MD, Director Neurosurgery UCLA, Los Angeles, CA; Dr. Luc D. Jasmin, MD, Neurosurgeon and Pain Management, Beverly Hills, CA; Dr. Louis T. Litsas, PA(ASCP) and Dr. Serguei I. Bannykh, MD, Pathologist, Cedars Sinai Medical Center, Los Angeles, CA; Mr. Michael DiMatteo and Ms. Patricia Conner Fredricks, DiMatteo & Associates, California Licensed Private Investigator, Moreno Valley, CA for all of their contributive efforts in the brain-chip mesogen specimen; Radiology Departments of University of Michigan, Ann Arbor, MI, and UCLA, Los Angeles, CA.

Introduction

Our world is made up of various elements, as you'll remember if you ever sat and stared at the periodic table in science class for many hours, while trying to understand the utilization of elements into compounds and their "3-D" structural composition. These elements are now used at 50 atoms units in the "nano world" to make the individual building blocks of nanotechnology. Nanotubes, nano-spacers, nano-rods, nano-wires and many other singlecomponents are even utilized to make the nano-radio, as used in nano medicine. We are a carbon friendly world which forms very strong covalent bonds; bonds in which atoms share electrons with each other, such as in the 3-D crystalline lattice of a diamond. Now, these elemental ingredients are in the laboratory's cauldron of nano ingredients to make different forms of mesogens. Mesogens are liquid crystals that have not only a specific smart function, but are used as the platform in developing new thin film transmitting nano ultra thin film coatings for biological sensors that have been used in "Brain Chip" mesogenic bio-sensors.

William E. Halal, Emerging Technologies and the Global Crisis of Maturity said,

> *"Although technological powers will be vast and progress will likely be made, the normal level of social resistance and political statement is likely to oppose change. Thus it may take an occasional environmental collapse, global wars and terrorist, or get unknown calamities to force the move to global consciousness."*

Even the NASA inventor and author, Ray Kurzweil in his book, The Singularity is Near, stated that by 2035, an implantable information chip could be developed and wired directly to the user's brain. The reality is that it is happening now through the advancements of sciences, computers, semiconductors, engineering, medicine, chemistry, and many other disciplines of science and engineering.

This text will address a specific case study of the removal of a menginoma, non-cancerous brain tumor, which in reality turned out to be a reactive nematic mesogen composed of advanced nano material brain chip/sensor for brain-computer interface. Additional mesogens will be discussed that were isolated from various human body surfaces and, specifically, from three different individual nasal passages, within a week of each other, that crossed the United States and Saudi Arabia.

Mesogens: Their Use in Building Nano-Biosensor Brain Chips

Mesogens are any compound that can exist as a mesophase and that part of the molecule of a liquid crystal that is responsible for its particular properties. In medicine, the term "mesogenic" pertains to, or is relating to, the capacity of a virus to lethally infect embryonic hosts after a short incubation period, but is incapable of infecting immature and adult hosts. In other words, the mesogen has a host-parasitic relationship in which the parasite dominates, but the host usually survives.[1, 2] This statement is right on target for how the mesogenic liquid crystal phase, and its other nano architectural designed smart functions, would interact with the biological system of the human brain or other target organs of designed choice. Mesogenic payloads of liquid crystals can be unreactive until triggered into activation for their ultimate purpose. This is the new nano medicine, new smart WIFI communication system, weapon, industrial espionage device, or component for mind or body control devices. In the nanotech world, there are no limits. It is the same with various types of mesogens. (See Figure 1)

To understand the technology of mesogens and their multiple smart functional uses for sensing, payload delivery systems, and other related aspects of their design, one must always remember that the mesogen is the tool or device – like the wheel on a car for the nano machine. The nano machine can be designed to become the robot, surgical scalpel, camera, the voice within your skull, or even the "false" collective consciousness when integrated into a computer system for collective mind control or a "global brain."

Types of Mesogens and Their Liquid Crystal Phases

A mesogen is primarily composed of liquid crystalline segmented copolymers systems that are composed of thin films of isotactic compounds that contain nano wires, tubes and other similar materials. They can function as sensors, transducers, actuators, and other related devices that are utilized in neurological mechanical device application or internal biological monitoring system.

The side chain of the liquid crystalline compounds create new doses of materials that offer the potential to couple the response of liquid crystals and elastomeric networks to applied mechanical strains. In the activation and design of functional strength and/or size parameter - such as pH, halogens, frequency (type) - hard and soft continuous domains and solvent solubility play vital roles in its design from nano material building units into an actual functional nano machine.

While liquid crystals are a state of matter that have properties between those of conventional liquid and those of a solid crystal, they can be divided into thermotrophic, lyotrophic, and metallotropic phases. Thermotrophic and lyotrophic LCs consist of organic molecules. Thermotrophic LCs exhibit a phase transition into the LC phase as temperature is changed. Lyotropic LCs exhibit phase transitions as a function of both temperature and concentration of the LC molecules in a solvent (typically water).

Metallotropic LCs are composed of both organic and inorganic molecules. Their LC transition depends not only on temperature and concentration, but also on the inorganic-organic composition ration. Liquid crystals can be found both in the natural world and in technological applications. Most modern electronic displays are liquid crystal based. Lyotropic liquid crystalline phases are abundant in living systems. For example, many proteins and cell membranes are LCs. Other well-known LC examples are solutions of soap and various related detergents, as well as the tobacco mosaic virus; all of which have been made into viral crystal protein envelops.

Liquid Crystal Phases

Liquid crystals were first examined in 1888 by Austrian botanical physiologist Friedrich Reinitzer, working at the *Charles University* in Prague, as he examined the physic-chemical properties of various derivatives of cholesterol, which are now known as cholesteric liquid crystals. Previously, other researchers had observed distinct color effects when cooling cholesterol derivatives just above the freezing point, but had not associated it with a new phenomenon. Reinitzer perceived that color changes in a derivative cholesteryl benzoate were not the most peculiar feature. He found that cholesteryl benzoate does not melt in the same manner as other compounds, but has low melting points. At 145.5 degrees Celsius it melts into a cloudy liquid, and at 178.5 degrees Celsius, it melts again and the cloudy liquid becomes clear. The phenomenon is reversible. This intermediate cloudy fluid with crystallites was called "liquid crystals."[3a, 3b] (See Table 1)

The various liquid crystal phases (called mesophases) can be characterized by the type of ordering. One can distinguish positional order (whether molecules are arranged in any sort of ordered lattice) and orientational order (whether molecules are mostly pointing in the same direction) and, moreover, order can be either short-range (only between molecules close to each other) or long-range (extending to larger, sometimes macroscopic, dimensions). Most thermotrophic liquid crystals will have an isotropic phase at high temperature, whereby heating will eventually drive them into a conventional liquid phase characterized by random and isotropic molecular ordering (little to no long-range order) and fluid-like flow behavior. Under other conditions, such as lower temperatures, an LC might inhabit one or more phases with significant anisotropic orientational structure and shortrange orientational order while still having an ability to flow.

The ordering of liquid crystalline phases is extensive on the molecular scale. This order extends up to the entire domain size, which may be on the order of micrometers, but usually does not extend to the macroscopic scale as often occurs in classical crystalline solids. However, some techniques such as the use of boundaries or an applied electric field (use of nano piezoelectrical materials) can be used to enforce a single ordered domain in a macroscopic liquid crystal sample. The ordering in a liquid crystal might extend along only one dimension, with the material being essentially disordered in the other two directions.

Liquid Crystal Elastomers (LCE)

Liquid crystal elastomers are dependent upon connectivity and conformation. Two polymer network conformations currently exist for LCE materials: Side chain LCE (SCLCE) and Main chain LCE (MCLCE) as illustrated in *Figure 1*. The side chain conformation has the LC mesogen pendant form the linear polymer network. The main chain LCE has the LC mesogens linked end to end forming a LC polymer which is cross-linked to form the network. Either nematic or smectic LCE may be prepared from side chain and main chain LCE. Some examples are the following:

- Acrylate side chain LCE - side chain LCE prepared through photopolymerization of acrylate mesogens. *(See Figure 2)*
- Siloxane side chain LCE - side chain LCE prepared via hydrosilylation reaction of vinyl mesogen with methyl-hydrosiloxane polymer. Materials of this type also include LCE films/ultra thin films (coatings).3 *(See Figure 3)*
- Main chain LCE - main chain LCE prepared via hydrosilylation reaction of vinyl mesogen with methyl hydrosiloxane polymer. *(See Figure 4)*
- Polyester/polyurethane (Estane) side chain LCE - side chain LCE segmented polyurethane with functionalized polysiloxane soft segments and tradiotnal MDI/butane diol hard segment. *(See Figure 5)*

Discotic Liquid Crystals (DLC)

Discotic liquid crystals, which are generally denoted by their aromatic mesogens, take advantage of the proximity of the aromatic mesogen to allow charge transfer in the stacking direction through the "pi" conjugate systems. The charge transfer allows the discotic liquid crystals to be electrically semi-conductive along the stacking direction. Applications have been focusing on using these systems in photovoltaic devices, organic light emitting diodes (OLED) and molecular wires (nanowires).[4, 5, 6] Discotics have also been suggested for use in compensation films, for LCD displays.

Photovoltaic devices - are discotic liquid crystals that have similar potential to the conducting polymers for their use in photovoltaic cells. They have the same technical challenges of low conductivity and sensitivity to UV damage as the polymer designs. However, one advantage is the self-healing properties of the discotic mesogens,[7] so far as the photovoltaic applications have been limited, using perylene and hexabenzocoronene mesogens in simple two-layer systems resulted in ~2% power efficiency.[8]

Organic light emitting diodes – So far, the study into discotic liquid crystals for light emitting diodes are at its infancy, but there have been some examples produced: A triphenylene and perylene-mesogen combination can be used to make a red LED.[9] The self-assembly properties make them more desirable for manufacturing purposes when producing commercial electronics than the currently used small molecule crystals, as in new commercial OLED displays. Also, they have the added benefit of self-healing properties that both the small molecule and the polymers lack as conductors, potentially being beneficial for longevity OLED products.

Thermotrophic Liquid Crystals (TLC) or Thermotrophic Crystal

Thermotrophic phases are those that occur in a certain temperature range. If the temperature rise is too high, thermal motion will destroy the delicate cooperative ordering of the LC phase, pushing the material into a conventional isotropic liquid phase. At too low a temperature, most LC materials will form a conventional crystal.[10] Many thermotrophic LCs exhibit a variety of phases as temperature is changed. For instance, a particular type of LC molecule (called mesogen) may exhibit various smectic and nematic (and finally isotropic) phases as temperature is increased. An example of a compound displaying thermotrophic LC behavior is paraazooxyanizole.

Nematic Phase or Biaxial nematic and Twisted Neumatic Field Effect

One of the most common LC phases is the nematic. The word *nematic* comes from the Greek *nema*, which means "thread." This term originates from the thread-like topological defects observed in nematics, which are formally called "disclinations." Nematics also exhibit so-called hedgehog topological defects. In a nematic phase, the calamitic or rod-shaped organic molecules have no positional order, but they self-align to have long range directional order with their long axes roughly parallel. Thus, the molecules are free to flow and their center of mass positions are randomly distributed as in a liquid, but still maintain their long-range directional order. Most nematics are uniaxials, meaning they have one axis that is longer and preferred with the other two being equivalent (can be approximated as cylinders or rods). However, some liquid crystals are biaxial nematics, meaning that in addition to orienting their long axis, they also went along a secondary axis. Nematics have fluidity similar to that of ordinary (isotropic) liquids, but they can easily be aligned by an external magnetic or electric field. Aligned nematics have the optical properties of uniaxial crystals and this makes them extremely useful in liquid crystal displays (LCD).

Smectic Phases

The smectic phases, which are found at lower temperatures than the nematic, form well defined layers that can slide from one another in a manner similar to that of soap. The smectics are thus postionally ordered along one direction. In the Smectic A phase, the molecules are oriented along the layer normal, while in the smectic C phase they are tilted away from the layer normal (*See Figures 6 A and B*). These phases are liquid-like within the layers. There are many different smectic phases, all characterized by different types and degrees of positional and orientational order.[11, 12]

Chiral Phases

The chiral nematic phase exhibits chirality (hardness). This phase is often called the cholesteric phase because it was first observed for cholesterol derivatives. Only chiral molecules (i.e. those that lack inversion symmetry) can give rise to such a phase. This phase exhibits a twisting of the molecules perpendicular to the direction with the molecular axis parallel to the direction. The finite twist angle being adjacent molecules is due to their asymmetric packing, which results in longer-range chiral order. In the smectic C* phase (an asterick denotes a chiral phase), the molecules have positional ordering in a layered structure (as in the other smectic phases) with the molecules tilted by a finite angle with respect to the layer normal. The chirality induces a finite azimuthal twist from one layer to the next producing a spiral twisting of the molecular axis along the layer normal.[13] (See Figure 7 and 8)

The chiral pitch, p, refers to the distance over which the LC molecules undergo a full 360 degree twist (but note that the structure of the chiral nematic phase repeats itself every half-pitch, since in this phase directors at 0 degrees and +/- 180 degrees are equivalent). The pitch, p, typically changes when the temperature is altered or when other molecules are added to the LC host (an achiral LC host material will form a chiral phase if doped with a chiral material), allowing the pitch of a given material to be tuned accordingly. In some liquid crystal systems, the pitch is of the same order as the wavelength of visible light. This causes these systems to exhibit unique optical properties - such as Bragg reflection and low-threshold laser emission[14] - and these properties are exploited in a number of optical applications. For the case of Bragg reflection, only the lowest-order reflection is allowed if the light is incident along the helical axis; whereas, for oblique incidence, higher-order reflections become permitted. Cholesteric liquid crystals also exhibit the unique property that they reflect circularly polarized light when it is incident along the helical axis and elliptically polarized if it comes in obliquely.[15]

Blue Phases

Blue phases are liquid crystal phases that appear in the temperature range between a chiral nematic phase and an isotropic liquid phase. Blue phases have a regular three-dimensional cubic structure of defects with lattice periods of several hundred nanometers; thus, they exhibit selective Bragg reflections in the wavelength range of visible light corresponding to the cubic lattice. It was theoretically predicted in 1981 that these phases can possess icosahedral symmetry similar to quasicrystals.[16, 17]

Although blue phases are of interest for fast light modulators or tunable photonic crystals, they exist in a very narrow temperature range, usually less than a few Kelvin. Stabilized at room temperature, blue phases allow electro-optical switching with response times of the order of 10_{-4} seconds.

Discotic Phases

Disk-shaped LC molecules can orient themselves in a layer-like fashion known as the discotic nematic phase. If the disks pack into stacks, the phase is called a discotic columnar. The columns themselves may be organized into rectangular or hexagonal arrays. Chiral discotic phases are similar to the chiral nematic phase. (See Photomicrograph 1)

Lyotropic Liquid Crystals

A lyotropic liquid crystal consists of two or more components that exhibit liquid-crystalline properties in certain concentration ranges. In the lyotropic phases, solvent molecules fill the space around the compounds to provide fluidity to the system. In contrast to thermotropic liquid crystals, these lyotropics have another degree of freedom of concentration that enables them to induce a variety of different phases.

A compound, which has two immiscible hydrophilic and hydrophobic parts within the same molecule, is called an amphiphilic molecule. Many amphiphilic molecules show lyotropic liquid-crystalline phase sequences depending on the volume balances between the hydrophilic part and the hydrophobic part. These structures are formed through the micro-phase segregation of two incompatible components on a nanometer scale. Soap is an everyday example of a lyotropic liquid crystal.[17, 18]

The content of water or other solvent molecules changes the self-assembled structures. At very low amphiphile concentration, the molecules will be dispersed randomly without any ordering. At slightly higher (but still low) concentration, amphiphilic molecules will spontaneously assemble into micelles or vesicles. This is done so as to "hide" the hydrophobic tail of the amphiphile inside the micelle core, exposing a hydrophilic (water-soluble) surface to aqueous solution. However, these spherical objects do not order themselves in solution. At higher concentrations, the assemblies will become ordered. A typical phase is a hexagonal columnar phase where the amphiphiles form long cylinders (again with a hydrophilic surface) that arrange themselves into, roughly, a hexagonal lattice. This is called the middle soap phase. At still higher concentration, a lamellar phase (near soap phase) may form, wherein extended sheets of amphiphiles are separated by thin layers of water. For some systems, a cubic (also called viscous isotropic) phase may exist between the hexagonal and lamellar phases, wherein spheres are formed that create a dense cubic lattice. These spheres may also be connected to one another, forming a bicutinuous cubic phase. (See Photomicrograph 2)

The objects created by amphiphiles are usually spherical (as in the case of micelles), but may also be disc-like (bicelles), rod-like, or biaxial (all three micelle axes are distinct). These anisotropic liquid crystals do form largescale versions of all the thermotropic phases (such as a neumatic phase of rod-shaped micelles). At high concentrations, inverse phases are observed; that is, one may generate an inverse hexagonal columnar phase (columns of water encapsulated by amphiphiles) or an inverse micellar phase (bulk liquid crystal sample with spherical water cavities. A generic progression of phases, going from low to high amphiphile concentration, is as follows:

□ Discontinuous cubic phase (micellar cubic phase)

- Hexagonal phase (hexagonal columnar phase) (middle phase)
- Lamellar phase
- Bicutaneous cubic phase
- Reverse Hexagonal Columnar phase
- Irreversible cubic phase (inverse micelluar phase)

Metallotropic Liquid Crystals

Liquid crystal phases can also be based on low-melting inorganic phases like $ZnCl_2$ that have a structure formed of linked tetrahedral and easily form glasses. The addition of long chain soap-like molecules leads to a series of new phases that show a variety of liquid crystalline behavior, as dual functions of the inorganic-organic composition ratio and of temperature. This class of material is known as metallotropic.

Biological Liquid Crystals

Lyotropic liquid-crystalline phases are abundant in living systems, the study of which is referred to as lipid polymorphism. Accordingly, lyotropic liquid crystals attract particular attention in the field of biomimetic chemistry. In particular, biological membranes and cell membranes are a form of liquid crystal. Their constituent molecules (phospholipids, acetylcholinesterase, enzymes, amino acids and other similar molecules) are perpendicular to the membrane surface, yet the membrane is flexible. The lipids vary in shape and can intermingle easily, but tend not to leave the membrane due to the high energy requirement of this process. Lipid molecules can flip from one side of the membrane to the other, a process catalyzed by flippases and floppases (depending on the direction of movement). These liquid crystal membrane phases can also host important proteins, such as receptors freely "floating" inside or partly outside the membrane.

Anisotropy of liquid crystals is a property not observed in other fluids. This anisotropy causes the flows of liquid crystals to behave differently than those of ordinary fluids. For example, injection of a flux of a liquid crystal - between two close parallel plates (viscous fingering) - causes orientation of the molecules to couple with the flow, with the resulting emergence of dendritic patterns. This anisotrophy is also manifested in the interfacial energy (surface tension) between different liquid crystal phases and it determines the equilibrium shape at the coexistence temperature. It is so strong, facets usually appear. When temperature is changed, one of the phases grows which forms different morphologies depending on the actual temperature change. Since growth is controlled by heat diffusion, anisotropy in thermal conductivity favors growth in specific directions, which has also an effect on the final shape.

Defining Nano Advanced Materials and Their Use as a Biosensor (Brain Chip)

Many of us in the fields of Health, Safety, and Environmental Engineering have recognized complex mixtures that form when chemicals combine with one another, which can create extremely hazardous situations. These materials are usually in the form of a solid, liquid, gas, or may be frozen as in a chryogenic substance. The difference with the nanostructures is based on their synthesis into a one-dimensional, two-dimensional, and now new three-dimensional forms. They use the same chemicals, but are now smart - like the Scarecrow in the Wizard of Oz movie who only wanted a brain. They have a brain because they can interface with computers through the transitional and piezoelectrial metallic materials they are made from. They can be a gel, particle, tube, claw, hook, fiber, or even a wire – but all exist under the penumbra of NANO.

In very recent press releases, nanowires are now being used in brain chips for the human brain so that a person can read the thoughts of another or transfer his thoughts to them. It is even being used for instantaneous language learning and retention. Nanowires are particularly interesting as they offer the opportunity to investigate electrical and thermal transport processes in size-confined systems (like the brain, human body, or even a plant) with the possibility of providing a deeper understanding of physics and other related sciences at a nanoscale. Silicon and silica nanostructures have attracted considerable attention because of their potential application in mesoscopic research and the development of nano-devices (tools and machines). The potential use of large surface area structures for catalysis or piezoelectrical thin film nanotechnology will be a whole new arena.

The growth of nano silica structures initiates from nanofibers composed of tiny amorphous particles. Aligned fiber arrays appear to grow from a single or biaxil nanofiber and have a structure similar to a protozoa and its "flagellum" (*See Figure 9 A*). The width of the bundle is 300-500 nm. Wang, *et al*, in 2000 observed that, after reaching a certain length, the silica fibers in the interior of the bundle cease growth while those in the outer regions continue to grow, forming a cylindrical chamber. Energy dispersive X-ray spectroscopy reveals small crystalline Si nanoparticles impregnated in the silica cylinder. Further growth of the nanofibers forms larger silicon (EDS) wire-like structure. The wire-like Si grows along the direction. In some cases, Si nanowires sheathed by a silica layer can be extensive as the growth direction remains (*See Figures 9 A, B and C*). The diameter of the Si core is -60 nm while the thickness of the silica sheath is -20 nm.

Silica wires can also be made to form a variety of unique three-dimensional structures. The nanofibers grow into bundles while paralleling a structure that has cylindrical symmetry. These arrays clearly demonstrate the versatility of the silica nanostructures which can be synthesized into specific shapes, like a "Chinese lantern" - a structure composed of Si and SiO_4 where an SiO_x tube extends from the top of a silica wire bundle. (*See Figure 10*)

The nanostructures demonstrated in this paper are dominant silica forms. Silicon nanocrystals/nanowires are formed in regions that are densely enclosed by the silica nano fibers. The short segment of the Si wire-like structure in *Figure 9A* is a typical example. This suggests that the growth may be dominated by vapor phase infiltration. In the regions dominated by silica, the local porous structure would appear to permit the diffusion of oxygen and silicon atoms through the materials, resulting in the growth of the silica structure. In the regions densely surrounded by silica, the diffusion of oxygen into the structure may be hindered resulting in the accumulation of silicon atoms and the formation of silicon nanostructures.

What is a Plain, Nematic or Reactive Mesogen's Role as a Bio-Sensor?

Newer developments in nanotechnology are now producing not just nanostructures, but taking these structures as their building materials to make nano tools and machines or functionally smart materials, such as mesongens. The chemicals used to make these types of materials consist of silica, silane, siloxane, acetone, ethanol, polypropylene, toluene, phenol, methylmethacrylate, polyurethane (as waterborne polyurethane), MDI/butane diol, and many others.

The polyurethane under scientific investigation for mesogens is a segmented copolymer with hard segments of MDI/butane diol. The soft segments are polysiloxanes with a cyano-biphenyl mesogen attached to each repeat unit via an eight-methylene spacer. The soft polysiloxane backbone ensures that the liquid crystal is free to move in this low temperature environment. Liquid crystalline mobility is further enhanced by the choice of the long spacer used to decouple the mesogen from the polymer main chain. The structure of this macromolecule, as well as a predicted morphology, is presented in *Figure 1*. The hard segments form hard domains ($T_{g\ g}$ = 88°C), and the soft continuous domains are in the smectic A phase over the entire liquid crystalline temperature range (T_g = -4.7°C); $T_{clearing}$ = 104.2°C).

Changes in these structures are dependent upon their nanostructure materials reaction to temperature, pH, halogens, oxygen, and frequency. Mesogens may be two-dimensional and then 8 layers or more high, just as you would build a high rise apartment complex. The nano term for this is "nematic," as in rising up. A reactive mesogen will react to temperature, pH, halogens, oxygen, and frequency. Frequency is very important because specific frequencies may be used to cause it to react, make it a specific structural morphology, trigger it to react and send an electrical impulse, or monitor its terrain (land, air, sea, or even a human body).

The overall synthetic design of the material is intended to provide a phase segregated morphology such that the pendant liquid crystalline groups in the soft segment would be free to respond to applied fields while anchored by the hard domains. A siloxane backbone may be used in the soft segment because siloxanes are known to be very low glass transition (T_g) materials. For instance, non-substituted hydromethylsiloxanes have T_g values on the order of -110°C. Siloxane oligomers substituted with the cyano-biphenyl mesogens of the type reported in this paper have been shown to have T_g values between -16°C and 25°C, depending on the length of the spacer group. [19]

The choice of the cyano-biphenyl mesogen is governed by the well characterized nature of this mesogen; and the cyano group is known to induce pneumatic liquid crystalline behavior. The relative stability of the neumatic versus smectic phases is controlled by the length of the alkyl spacer, which favors smectic phases as its length increases. Another advantage of the mesogen is the clear signature that the cyano group affords in infrared spectroscopic characterizations. This identification tag allows the infrared dichroism studies to reveal the orientation of the mesogen on the applications of a stress field.[20a]

Specimens collected in June 2010 - from Los Angeles, New York City and Saudi Arabia of nematic mesogens from the nasal bulb of 3 separate individuals - showed in analytical testing that they were made out of the same materials and that the nano claw, or hooks, were all composed of "super glue" upon Raman/Micro Fourier Trans Infrared Spectroscopy analysis.[20b] (See Photomicrograph 3 and Figure 11).

HUMAN BRAIN CHIP CASE STUDY: From Exposure to Device

In July 2011, *Integrative Health Systems®, LLC* (IHS) received a referral patient for comprehensive industrial toxicology evaluation and advanced biological monitoring testing. During that time, the patient explained how she was exposed to multiple directed energy sources that were monitored within her home. The individual, a female of Chinese ancestry, age 57, a professional cultural educational director/musician, explained that approximately two years prior she was sitting at her desk talking on the phone (right ear) when she experienced a feeling of burning-glass needles entering into her left ear and its canal. Several weeks later, she started to experience pains on the left lobe area of her brain and also experienced a cough that came from the upper bronchi area of the left lung. Testing performed by her physicians revealed - in 2010 and reconfirmed in 2011 - a 9 mm menginoma on the left lobe of the brain. The patient/client was determined to remove the menginoma and to have it tested, so surgery was scheduled at Cedar Sinai Hospital with the surgeon being UCLA's Director of Neurosurgery. The brain specimen was collected by a private investigator retained by the client/patient and the ear canal specimen was transported to *Integrative Health Systems®, LLC* by the client/patient. All appropriate chain of custody procedures were followed.

It should be mentioned that a 1.2 cm menginoma was removed from a larger 9 mm menginoma and was forwarded to *Integrative Health Systems®, LLC* for advanced nanomaterials comparative analysis, as originally seen in the two MRI's.[21] The results are stated in this paper as in the testing performed; not only on the brain menginoma, but the two ear canal specimens which were washed out of the ear canal on a separate date. All specimens went through the following phases of analysis as stated below:

- Phase I: Photomicrograph of Specimen
- Phase II: Energy Dispersive Spectroscopy/Scanning Electron Microscopy
- Phase III: Raman/Micro Fourier Trans Infrared analysis

(*See Photomicrograph 3 and 4*)

Discussion, Findings and Conclusions of Brain Menginoma Results

Phase III: Raman/Micro FTIR (Fourier Trans Infrared) analysis was performed on the sphere inside and out of the sphere composition (the outside layer and inner core area). In addition, both areas were re-tested for specific locations associated with materials observed on the outside of the sphere, sliced open, and then the inside core. Note that on each EDS analysis there was reddish material thought to be human blood, but no iron was present in the analysis. Iron must be present in blood to be human blood. This specimen could not be cut with a scalpel. A diamond cutter had to be used on the specimen.

The following results and discussion is based upon Specimen L/N 19666, Brain Mengionoma, collected November 7, 2011 from *Cedar Sinai Hospital*. This particular section will address the mesogen sphere found within the left lobe of the brain.

Outer Whole Piece Layer, inside (cut), EDS/SEM Analysis: 20.165 % Carbon; 42.257 % Oxygen; 0.650 % Sodium; 0.289 % Magnesium; 0.229 % Silicon; 10.198 % Phosphorus; 0.435 % Chlorine; 25.778 % Calcium. (Note: Pinkish colored piece - with trace of blood on the outside surface - was not blood because no iron was present, which is required for blood.)
Micro FTIR analysis result was hydrolyzed animal protein. Same specimen, but the purple area which will be called Sample # 3: EDS of purple area contained: carbon, oxygen, sodium, silicon, sulfur, chlorine, potassium, and phosphorous.

Micro FTIR analysis result was outside surface hydroxyapatite, calcium carbonate, and animal protein.

Micro FTIR analysis result for inside surface was animal protein.

The melting point was initially >300 degrees Celsius. A more in-depth melting point test was done and its findings are reflected in *Table 1*. The final result being that the material went through multiple stages of change at various temperatures and ended up not melting at >1100 degrees Celsius and morphologically developed into green powdered crystals.

Brain Menginoma Relative Findings

The outer whitish material under Micro FTIR analysis, by utilization of Bio-Rad equipment, was pentacalciumhydroxyphosphate and hydroxylapatite. The outer region was calcium carbonate and animal protein. The animal protein had some additional unique peaks on the chromatograph which may be due to substance amplification with specific charged elements, as in thin film nano technology coatings. (*See Photomicrograph 12*)

The melting point was >300 degrees C with no other discussion in the report, such as if it burned up or remained the same. Note that Human enamel has a melting point of 1268 degrees C. The melting points do not match.

Hydroxyapatite is also known as pentacalcium hydroxyphosphate. In normal biological conditions, it is known as "brain sand" or corpora arenacea which can calcify within the brain - especially within the pineal gland area. These materials are used as fillers to replace amputated bone, or as a coating to promote bone growth into prosthetic implants. This material is very porous in its natural and synthetic forms, which allows other nano composite materials to adhere to it in layers. It is used to stabilize strontium and other trace minerals as it would be used to determine collagen content in bone.

Hydroxyapatite (Hap) is a major inorganic component of human hard tissue, such as bones and teeth, and its content determines their microstructures and physical properties. Artificial Hap shows strong biocompatibility and bioactivity and has found broad applications in tissue engineering for replacing damaged hard tissue. However, the artificial Hap suffers from its intrinsic low mechanical properties. So, to meet mechanical requirements, Hap can be incorporated with stiff mineral phases (mullite, zirconia, alumina). The performance and long term survival of these biomedical devices are also dependent on the presence of bacteria surrounding the implants. In order to reduce the incidence of implant associated infections, several treatments have been proposed, such as the introduction of silver or fluoride in the Hap. The objective of particular research conducted by the University of Pisa in Pisa, Italy, by B. Cioni, *et al*, is the sintering of composites based on calcium phosphate; mainly, Hap supported zirconia for bone replacement with better microstructural features. In fact, the use of zirconia can enhance the mechanical properties of bio-ceramics. Moreover, the introduction of small amounts of silver, which should improve the antibacterial properties, has been taken into consideration. Note that silicon was present, which is also found in zirconia complexing agents with metal oxides.

The hydroxyapatite found in this specimen is a biomimetic synthetic material that is used to transfer information and, consequently, to act selectively in the biological environment.

Synthetic biomaterials can be turned biomimetic by imprinting them with the morphology of biogenic materials. Biomimetic hydroxyapatite represents an elective material because it has a very similar composition to the mineral component of bone; and, moreover, its chemical-physical properties and surface reactivity can be managed by modifying synthetic parameters. The morphology of synthetic biomimetic hydroxyapatite is essential to optimize its interaction with biological tissues and to mimic biogenic materials in their functionalities.

Biomineral morphogenesis is related to specific strategies for the long-range chemical construction of well organized architectures from preformed nano or micro crystalline inorganic building blocks. In fact, many biologic complex structures are obtained by promoting specific links induced by the conformation variability at the nanometric scale of the biological macromolecules. Vertebrate bones and teeth are biological hybrid materials where a calcium phosphate, in the form of Hydroxyapatite (HA), represents the inorganic component intimately inter grown with the organic matter prevalently constituted of proteins and polysaccahrides. To accomplish this technique, the physical-chemical features should be tailored in synthetic biomimetic HA nanocrystals that are the dimensions, porosity, morphology and surface properties. The surface functionalization of HA nano-crystals with bioactive molecules makes them able to transfer information to, and to act selectivity, on the biological environment. This represents a primary challenge in the innovative bone substitute materials into the biosensory applied devices for advanced nano materials.

The presence of calcium carbonate, in addition to hydroxyapatite, suggests that the initial materials which entered the left ear canal were primarily composed of carbonate-hydroxyapatite nanocrystals (CHA) that have been synthesized with a nearly stoichiometric in bulk Ca/P molar ratio of about 1.6-1.7, containing 4 plus/minus wt % of carbonate ions replacing prevalent phosphate groups. If you look at the EDS results, this range does appear to be the same for Specimen LN 19666.[23, 24, 25] (See Figure 13)

Phase I: Microphotograph of the sphere and cut sphere showed a reactive nematic mesogen. (See Photomicrograph 4 and 5) It has been determined that this reactive nematic mesogen may be part of a nano tool/device known as a liquid-crystalline phase development of a mesogen-jacket. The liquid crystalline phase is stimulated by attenuated frequency and pH dispersion. These substances are composed of silicon-stabilized tri-calcium phosphate, both of which were found in EDS analysis. Due to the ratio of calcium to phosphate, it is possible the tri-calcium phosphate molecules were originally present. The decrease in the hydroxyapatite content is proportional to the "drug delivery system" design of this type of material.

Hydroxyapatite is delivered to expand in the mesogen jacketed material via the composition of the metal oxide thin films and other compounds, such as acetycholinesterase chloride or heprin. Coatings of nano calcium phosphate, as used in biomimetic synthetic materials or MEMS, are for antibiotic purposes. With the presence of strontium silicon or zirconium compounds, there will be a depreciation of biological Vitamin D within human subjects due to strontium inhibiting the metabolites of Vitamin D (OH, 1,25).

Brain Menginoma Conclusions

In advanced biological monitoring tests performed on patient/client, there were elevated strontium, arsenic, nickel, and acetylcholinesterase as well as depleted Vitamin D. To deliver the synthetic calcium hydroxyapatite nano crystals, as utilized in a mesogen jacketed process of delivery, there would have to be the presence of acetylcholinesterse chloride (as in nano tube composition) and/or heprin. The patient/client did have elevated values of

16,005 IU acetylcholinestease, RBC with occupational levels of nickel and metabolites of styrene as found in polyurethane epoxies or polymers.

In many of the designs of this type of biomaterial, there may be a DNA tag or sensor utilizing a polyamide or polyurethane. These types of materials are utilized to monitor DNA as a biosensor and/or show crystal growth behavior to utilize a one-dimensional copper core with a nickel shell. Nickel was found at high levels within patient/client's blood.

If the materials are being utilized for monitoring Nano/Info/Bio-Tech through the monitoring of brain functions, then the material may be registered under studies being conducted by the AMPTIAC sponsored information analysis center of the US Department of Defense (DOD). Some of these types of brain or CNS bio-sensory research applications have been granted to *US Nanocorp, Inc.*, *University of Connecticut*, *Lifeboat Foundation*, Dr. David E. Reisner, *Rice University* and its applications for nanodielectric materials for high energy density capacitors. Under US Military applications, there have been projects that deal with specific tasks of textiles, biosensors, and fluidic lab-on a chip MEMS under small business developmental grants for the *US Army*, *US Air Force* and (1991-1995) by the *Naval Air Warfare Center* in China Lake, CA.[27, 28, 29, 30]

The material found to compose the reactive nematic mesogen with mesogen jacketed calcium hydroxyapatite nano crystal scaffolding is a synthetic manmade material. It is used to monitor the brain, CNS, and other biological activities of the body. Current analysis and patent information appear to show that this is made in the USA, with possible investment aspects in manufacturing or intellectual technology bridges into Hong Kong, China, Korea and Japan. [31] Additional confirmation will be revealed after obtaining the Micro FTIR results of the ear canal calcification specimen, which will be addressed in a separate section of this text. These materials may also be composed of terahertz photonic crystals that have a calcium carbonate and an organic polymer mesogen epoxy, such as of a urethane base.

Ear Canal Specimen Findings, Results and Conclusions

The two ear canal specimens are identified as *19655 Sample A* and *19655 Sample B*. Phase I – Microphotographs, Phase II - EDS/SEM analysis, and Phase III - Raman/Micro FTIR analysis were performed on each of the specimens. In addition, a microorganism analysis was performed on the water used to wash out the materials from the ear, by the patient/client's physician. Specimens were collected by the patient/client and hand delivered to *Integrative Health Systems, LLC* for advanced nano material analysis and specific collaborative research auspices. (*See Photomicrograph 5*)

Ear Canal EDS/SEM Results *(See Figure 14)*

19655 Sample A: 56.654% Carbon; 38.591% Oxygen; 0.023% Silicon; 1.447% Sulfur; 0.670% Chloride; 0.970% Potassium; and 1.645% Calcium.
19655 Sample B: 56.847% Carbon; 41.614% Oxygen; 0.668% Sulfur; 0.163% Chloride; 0.267% Potassium; and, 0.440% Calcium.
19655 Sample B Fiber: 52.046% Carbon; 46.158% Oxygen; 1.335% Sulfur; 0.061% Chlorine; 0.113% Potassium; and 0.287% Calcium.

Ear Canal RAMAN/Micro FTIR Results *(See Figure 15 and 16)*

19655 1 B (large particle): Specific band assignments for protein bands are "animal protein". A detailed discussion will address the types of animal protein, which is not human protein.

3300 cm-1	NH stretch
3080 cm-1	CNH overtone
3000-2850 cm-1	CH stretch
1650 cm-1	C=O
1535 cm-1	CNH
1396 cm-1	CH rock
1493 cm-1	CH3
1240 cm-1	CNH
730 cm-1	NH wag

19655/2A and 19655/2B: 2A- Black Square particles featured small hydroxyl OH group.

19655/2A *black square particle (Raman)* compared to synthetic graphite standard (Alfa Aesar) green laser gave a comparison analysis match to graphite.

19655/2B (Body): matched to Moon Glycerine U.S.P. (x0.90) (See Figure 17)

19655/2B (Body): and Marvanol DP (See Figure 18)

Additional laboratory Bio-Rad analysis results and observation comments were the following:

- The sample 19655/2A black square particles did not produce a response using FTIR. It was analyzed using dispersive Raman spectroscopy with a 532 nm laser, and determined to be highly consistent with graphite.

- The sample 19655/2B body mass was determined to be a mixture of two compounds. One was highly consistent with glycerine; the other, a surfactant *Marvanol DP*. This portion of the spectrum is consistent with a long chain amide polymer.

DISCUSSION OF THE ADVANCED NANO MATERIAL ARCHITECTURE DESIGN

Advanced nano materials have specific architectural design as they are formulated - via their specific chemical composition - with a starting functional number of 50 atoms. This is important to note because the actual design of the specific nano material is then incorporated into a layered film, glue-composite, rod, wire, and other functional nano building blocks to create the specific thin film type, mesogen (layered 6 to 8 stories high when expanded) and specific nano devices. All these materials are man-made.
The delivery system of these materials may be through aerial dispersion - as in smart dust and smart crystal motes - and while other materials may not only need pH and halogens to react with the biological system they are found within, but frequency. The types of materials utilized in the chemical compounds to make the nano device into a liquid crystal, crystal mote, nanobot, and many other functional mechanical devices - that are specifically designed to integrate into the biological terrain - will always have a specific "smart function" behind their design.

The visual appearance of specimens through photomicrographs allows the observer to note specific design shapes, sub-parts, and other architectural design features. This is very important because one advanced nano material may very well be made up of multiple parts, chemicals, and the utilization of materials within the biological system to be used to form the device.

In the patient/client's case, she was sitting at her desk talking on the telephone with the phone receiver on her right ear. Approximately two years ago, she experienced a sharp burning glass-needle pain in her left ear that traveled down the ear canal. After that event, she experienced headaches and other symptoms related to fluctuations of high/low amounts of acetylcholinesterase enzyme activity. In 2010, the menginoma was observed on her left brain lobe by the *University of Michigan* at Ann Arbor. It was removed in 2011 by the Director of Neurosurgery at UCLA. The specimen was collected and sent to *Integrative Health Systems®, LLC,* Dr. Hildegarde Staninger®, RIET-1, for specific Advanced Nano Material research analysis. The result was a reactive nematic mesogen, which is a nano material device with specific design and function. Due to its chemical composition, it was determined to be a brain bio-sensor and/or related device.

The delivery system for this material was an injection into the ear canal. In many cases, an individual may be exposed to nano materials on separate occasions. This may be true in the patient/client's case, since she experienced a specific type of descriptive pain, which is continually the same for exposure to advanced nano materials "burning glass-needle." This particular experience can be addressed as a "nano needle delivery system" since the standard neurological response for exposure to nano needles is a feeling of a burning glass-needle on the surface of the skin. She did notice a movement in the bushes outside of her office window. To create a sound stream or laser pulse, one would use a hand held sound, microwave, or pulse laser to accomplish this task. The wave would have to penetrate through glass.

These facts are important since the material found in the Raman and Micro FTIR analysis of the two ear canal specimens were never exposed to any surfactants or ear wash solution except for water during the ear wash procedure performed by her physician.

The materials isolated in the ear canal pieces, as well as the ear "wax" specimen B, have very unique compositions that are composed of compounds utilized in the making of advanced nano materials. The patient/client's clinical biological monitoring tests had specific parameters present in high amounts, which are relative to the nano material compositions. The materials found were arsenic, nickel, and acetylcholinesterase. The mesogen from the brain contained hydroxypatite, which is very important for the types of protein bandwidths identified in Raman/Micro FTIR analysis results of the ear canal specimens. It must be remembered at all times that we are not dealing with a specific whole compound, but rather a composite of multiple compounds that comprise ultra thin films, wires, rods, etc. to make a functional nano or bio-nano material.

The Raman analysis of the black square particles showed no signal, which then determined the material to be graphite. When a Raman signal shows absolutely no signal, there can be the presence of "nickel." The nickel has a specific function of creating steam reforming of glycerol into hydrogen when done over nano-size, nickel-based, hydrotacite-like catalysts. Hydrotacite is also called hydrocarbonate where carbonates are present in the EDS/SEM analysis of both Sample A and B 19655 Ear Canal Specimens.

In the nickel nano particle process, steam (sweating) or evaporation from the ear canal or from steam reforming (SR) of glycerol for the production of hydrogen over the nano-sized Ni-based catalyst occurs. The Ni-based catalyst was prepared by solid phase crystallization and impregnation methods, as characterized by the ananlysis results in this paper. The Ni could have contained a Ni/gamma-Al2O3 catalyst which would show a higher conversion and H2 selectivity. When the catalyst is utilized, it would slowly deactivate due to the carbon formation (graphite) on the surface of the catalyst and then inserting it. In studies by E. Hur and DJ Moon in 2011, it was found that the nickel based hydrotalcite-like catalyst (spc-Ni/MgAl) showed higher catalytic activity to prevent carbon formation than Ni/gamma-Al2O3 catalyst in the SR of glycerol.

The tissue within the ear canal has a pH of 4.8 to 5.5 (acidic). Acid-basic materials are often used to catalyze organic reactions. Hydroxyapatite is acidic and hydrotalcite presents basic properties. The association of both compounds in a single material should present a rather unique catalytic a behavior. Studies conducted by Rivera, Fetter and Bosch in 2008[25] showed that when they made preparations of hydrooxyapatite impregnated with hydrotalcite (hydrocarbonate) and exposed it to microwave irradiation at three different preparation levels, the following results occurred:

- A homogeneous distribution of hydrotalcite on hydrooxyapatite surface is obtained when hydrotalcite is precipitated over a previously microwave irradiated hydrooxyapatite.

- Alternatively, if the hydrotactite mixture is incorporated to the hdyroxyapatite precursor gel and the resulting mixture is microwave irradiated, hydrotacite is preferentially deposited in the hdyrooxyapatite interparticle spaces. *(Note: Hydrooxyapatite was on the outside core of the mesogen from the brain with a core of animal protein. The gel may be composed of a silica sol-gel formulation by siloxane thin coatings.)*

- When both hydroxyapatite and hydrotalcite solutions are irradiated, mixed, then irradiated again, the composite behaves as the addition of the two components.

In the hard ear sample called *19655 A*, there was silicon present. Silicon is a mixture of silica, silicon dioxide, and quartz. The analysis showed carbon results, which can be from silicon carbide (SiC) and other related materials. Grains of silicon carbide (SiC) have been used in the formation extremely refractory and chemically resistant minerals, such as diamond, graphite, silicon carbide, silicon nitride, and various refractory oxides.[26,27,28]

Silicon carbide (SiC) can be dissolved in HCl/HF/HCL to dissolve silicates, CS_2 extraction of elemental sulfur, light oxidation using Na dichromate, colloidal extraction in ammonia of nanodiamonds, and cleaning of the diamond fraction by treatment with perchloric acid resulting in pure nanodiamond materials.[29,30,31] These materials are then mixed into sol-gel which are then formed into liquid crystals and nano composites with animal glue. Animal proteins do comprise the processes of glues when dealing with specific composite materials that will spread, anchor, and then form a specific combined material.[32,33]

The single compositional elements of sulfur, chloride, potassium, and calcium were found throughout both specimens as in association with the identification of glycerol, *Marvanol DP* (surfactant, salt), and an amine polymer. In nano-encapsulation of proteins via self-assembly with lipids and polymers,[33] the natural factor of squalene monooxygenase enzyme will be utilized by these materials. The specific range found on the Micro FTIR bandwidth was 3000-2850, which is for 2-amino benzhydrol template for highly reactive inhibitors utilized in squaline synthesizes of mesogenic biological and polymeric materials for ultra thin film nanotechnology.

Glycerols are lipids and, in a nanoscale device, may be used in self-assembly with branched polymer of polyethyleneimine (PEI) or other amphiphili poly (ethyleneglcol) monooleate derivatives or salts. The compound *Marvonal DP* is such a compound. The created self-assembly nanovehicles can bind to specific proteins in the presence of these nano-objects. The cationic polymer PEI formed mixed nano-objects with the protein. The polymer conformation in the nanovehicle will establish to sensitive dilution, a property that can be essential for the protein release upon administration, such as observed with animal protein and hydroxypatite. The amphiphile compounds will create small micelles in the absence of a lipid (the glycerols).[34a]

Further crystallization within a silicon carbide and related minerals can show evidence of nanometric inclusions and mineral association, which contains specific elements of Si, SiO2, TiO2, Fe, Ti, Cr, Zr, Ba, Mg, Th, P, S, Pb, Zn, Nb, Al Ca, and Cl.[35] These facts are restated due to the elements found on EDS/SEM results and in biological specimens of client. The nano diamond may have specific inclusions that other materials are attached to it so that it may deliver to other areas of the body as it is dispersed throughout its delivery area.

Graphite is used in the formation of nano carbon tubes and other materials, such as Bucky balls. Single layers of graphite were observed by Raman Spectroscopy and Scanning Transmission Electron Microscopy, within the ear canal specimen, as black square particles. *Manchester University* researchers in 2004[34b] pulled out graphene layers from graphite and transferred them onto thin SiO2 on silicon wafer in a process sometimes called micromechanical cleavage or, simply, the Scotch tape technique. The Si2 electrically isolated the graphene and was weakly interacted with the graphene, providing nearly charge-neutral graphene layers. The silicon beneath the SiO2 could be used as a "back gate" electrode to vary the charge density in the graphene layer over a wide range, which will lead into the quantum Hall effect in graphene.[36] More recently, graphene samples prepared on nickel films, as well as on both the silicon face and carbon face of silicon carbide, have shown the anomalous quantum Hall effect directly in electrical measurements.[36,37,38,39,40] Graphitic layers on the carbon face of silicon carbide show a clear Dirac spectrum in angle-resolved photoemission experiments, and the anomalous quantum Hall effect is observed in cyclotron resonance and tunneling experiments. Even though graphene on nickel and on silicon carbide have both existed in the laboratory for decades, it was graphene mechanically exfoliated on SiO2 that provided the first proof of the Dirac fermion nature of electrons in graphene.

Other Relative MicroFTIR Bandwidths Animal Proteins

~1650 cm-1: Cyanobacteria, Calotric sp. protein sheath, the whole-cell spectra are dominated by the amide I feature at ~ 1650-cm1. This band is sensitive to the protein's secondary structure, to ligand interactions, and to its folding characteristics.[41,42] In chemically purified sheath, the amide I and II bands are weaker and the amid II band has shifted to lower frequencies as a consequence of thermal denaturation due to the preparation. This is true for delivery into other biological terrains of proteins.

~1240 cm-1: Cyanobacteria, Calotric sp. nucleic acids corresponding to the asymmetric phosphodiester stretching vibration of the DNA or RNA backbone structure.[43,44]

~3500-3300 cm-1: Present in all spectra liquid or gaseous water and gaseous CO2 (at ~2360) are often present. The amount of water and the amount of water vapor in the area is directly constant upon dried specimens. The frequency bandwidths of the following animal proteins found in the ear canal specimens have a dual role in electronics, polymer-based widebandwidth, and other characteristic functions. The specific frequency will be addressed.

~730 cm-1: Material associated with polyethylene-polystyrene gradient polymers and their swelling abilities. Usually accompanied by a half-width of 720-730 cm-1 utilized in tromicrographs and micromachined tunneling accelerometers coupled with silicon for capacitive bulk acoustic wave silicon tracking.[45]

~1493 cm-1: PS spheres of PS colloidal crystals (PS – polystyrene) utilized in quantum wells as related to microhyration of stretching the water molecule via metastable and microparticulate hydrated materials.[46]

~1396 cm-1: Single crystal IR spectroscopy of very strong hydrogen bonds of pectolite.[22]

~1535 cm-1: Nano micro resonators made of Ir and porous SiO2 with Bragg mirrors consisting of alternating the layers for a tunable band pass fiber optic filter as associated with protein secondary structures.[47]

~3080 cm-1: Three-cavity tunable MEMs through IR light Fabry Perot Interfermoteric principle-optical filter. Specific band width of cloned ca3080 for carbonaceous materials to absorb amplitudes, such as the Samsung SND-3080 in the Samsung Techwin's elite line of IP Security computer generated enabling camera materials.[48]

Ear Canal Specimens Conclusion

The materials identified in the ear canal specimens are composed of advanced nano materials that are designed specifically to deliver a variable payload into directed areas of the biological terrain. In the patient/client's case, this was her brain, upper bronchial area, and other areas of the body. The materials are designed to form under specific frequency releases into working mechanical devices that may tunnel into the tissue or specific areas of the body.[49]

The mesogen associated with patient/client's menginoma was composed of materials that were delivered to the area (brain) via the original materials in the ear canal. Specific frequencies, as identified in Raman and Micro FTIR, have precise chemical components that were found in her body, such as metabolites for styrene, nickel, acetylcholinesterase, and porphyrins.

It is important to realize that many of the specific frequencies identified in the Micro FTIR identified materials are used to allow explicit tracking or biometric equipment to track, monitor, and control or take pictures utilizing the human subject as a walking-talking monitoring device. This process is highly complex and involves many levels of science, medicine, physics, and engineering. In such cases, the materials within the biological specimen will attract specific forms of energy and react. This means that no direct energy wave or emission from an outside device has to be present. It may be done on a computer system, smart phone, satellite, wireless phone tower, or other similar devices.[50,51,52]

Since the hydroxypatite is associated with payload delivery systems utilizing microwave, the original activation of her ear canal may have been from a microwave hand-held device or similar. The specific frequency for the materials at ~730 cm-1 bandwidth is for styrene, which is used for tunneling. The patient/client has perceived continual movement and can feel the materials moving. It should be remembered that many of these materials would have a metal or piezoelectric material associated with them to enhance their ability to bind, attach, or anchor to the nervous system as it would establish a micro array pattern or grid. The grid system would be to the core of the individual, i.e. bone.[53,54,55]

The use of cyanobacteria protein sheaths and DNA/RNA may be utilized as a specific material to set tracking devices to, or to program reactions. This fact is stated because it is very rare that cyanobacteria protein would be found in the human body; and, it was in the nano composite material that made up the ear canal specimen.

It must be emphasized that the Nano Molecular Identification and Nano Architecture of man-made mechanical devices is through a highly skilled approach in order to put all of the morphological characteristically facts together.

The original menginoma mesogenic sphere had further melting point analysis done on it, which is contained in the data of this report. It must be noted that the specimen continued to change in composition with higher temperature exposure. The final result was that it turned into a green crystalline powder at >1100 degrees Celsius. It should further be noted that graphite, a man-made compound, turns green at 1100 degrees Celsius in the presence of nickel.

The materials found in the ear canal specimens could have also been utilized as a sending and receiving device for acoustical transmissions when coupled with specific advanced materials.

CONCLUSIONS AND SUMMARY

The integration of nanotechnology into the bio-sensory world to monitor or control human life is where the line in the sand is drawn for the human being. The majority of High-Impact Technologies that utilize brain-computer interfaces as a "Brain Chip" or Bio-Sensor would be for the following:

- Control and monitoring of the brain and bodily functions.
- Control and monitoring of the behavior of the individual.
- Sending and receiving verbal commands.
- Stimulation of bio-electrical transmissions within the neuron trees of the nerves.
- To be utilized as a listening device for remote sensing and monitoring.
- To be used as a transmitter for listening in on conversations within a specific area that the individual may be in where the device has been implanted in them.
- If there is a digital computer component to the device, it could be used to capture visual transmissions as a walking/talking monitoring system (a high-tech extrinsic spying system, especially for industrial/military espionage).
- Many other aspects as related to the multiple use applications of mesogens.

It is our own individual divine right as a human being, created in God's own image by his own hands, to be a HUMAN. A human that has free will as originally designed by the Creator and not to be enslaved by a man-made "Brain Chip" that can be dispersed in the air, shot at you with a microwave, then controlled by your TV, computer, or cell phone. When these materials interface into the Human, they become a Transhuman controlled by another Transhuman - who may have very well been a Human at one time. NOW is the time for all of humanity to realize that enough is enough and no human was created to be turned into a "POST HUMAN SPECIES." That was never the true divine plan of the universe.

REFERENCES

1) US EPA. Recognition and Management of Pesticide Poisonings, 4th Edition. EPA 540/9-88-001, March 1989. US Government Printing Office, Washington, D.C., Chapter 1: Organophosphate Insecticides and Chapter 2: N-Methyl Carbamate Insecticides. Pages 1 -12

2) Gregoriou, Vasilis G. and Mark S. Braiman. Vibrational Spectroscopy of Biological and Polymeric Materials. CRC Press: Taylor & Francis. Boca Raton, FL ©2006

3) a. Kaloyanova, Fina P. and Mostafa A. El Batawi. Human Toxicology of Pesticides. CRC Press. Boca Raton, FL ©1991 b. Sulckin, TJ, Dunmur, DA and H. Stegemeyer. Crystals that Flow- Classic papers form the history of liquid crystals. London: Taylor and Francis. ISBN 0-415-25789-1 ©2004

4) Amdur, Mary O., Doull, John and Curtis D. Klaassen. Cassarett and Doull's Toxicology: The Basic Science of Poisons, 4th Edition. Pergamon Press. New York ©1991

5) Collings, Peter J. Liquid Crystals: Nature's Delicate Phase of Matter, 2nd Edition. Princeton University Press, Princeton ©2002 and Oxford ©1990. Oxfordshire, England

6) Institute of Medicine and National Research Council of the National Academies. Globalization, Biosecurity and the Future of the Life Sciences. The National Academies Press, Washington, D.C. ©2006

7) Gregoriou, Vasilis G. and Mark S. Braiman. Vibrational Spectroscopy of Biological and Polymeric Materials. CRC Taylor & Francis. Boca Raton, FL ©2006

8) Barbero, G. and L.R. Evangelista. An Elementary Course on the Continuum Theroy for Nematic Liquid Crystals. World Scientific Publishing Co., Pte. Ltd., Singapore ©2001

9) Collings, Peter J. Liquid Crystals: Nature's Delicate Phase of Matter, Second Edition. Princeton University Press, Princeton and Oxford. Oxford Shire, England. ©1990 (Oxford) & 2002 (Princeton).

10) National Research Council. Sensor Systems for Biological Agent Attacks: Protecting Buildings and Military Bases. National Academy Press. Washington, D.C. ©2005

11) Pollack, Gerald H. Cells, Gels and the Engines of Life: A New, Unifying Approach to Cell Function. Ebner and Sons Publishers. Seattle, Washington ©2001

12) Wiwanitkit, Viroj. Advanced Nanomedicine and Nanobiotechnology. Nova Science Publishers, Inc. New York, New York ©2008

13) Williams, Linda and Dr. Wade Adams. Nanotechnology Demystified A Self-Teaching Guide. The McGraw-Hill Companies. New York, New York ©2007

14) Booker, Richard and Earl Boysen. Nanotechnology for Dummies. Wiley Publishing, Inc. Indianopolis, Indiana ©2005

15) Freitas, Jr. Robert A. Nanomedicine: Volume 1: Basic Capabilities. Landes Bioscience. Austin, Texas. ©1999

16) Sacarello, Hildegarde. The Comprehensive Handbook of Hazardous Materials:

Regulation, Handling, Monitoring and Safety. Lewis Publishers/CRC Press, Inc. Boca Raton, FL ©1994

17) Boucher, Patrick M. Nanotechnology: Legal Aspects. CRC Press, Inc. Boca Raton, FL ©2008

18) Lin, Patrick; James Moor; and John Weckert. NANOETHICS: The Ethical and Social Implications of Nanotechnology. Wiley-Interscience (a John Wiley & Son, Inc. Publication). Hoboken, New Jersey ©2007

19) Wang, Shong I., Gao, Ruiping P. , Gole, James I. and John D. Stout. "*Silica Nanotubes and Nanofiber Arrays*". Advanced Materials © 200, 12, No. 24, December 15, pgs 1938-1940.

20) a. Executive Office of the President. National Science and Technology Council. National Nanotechnology Initiative Strategic Plan. National Science and Technology Council, Committee on Technology Subcommittee on Nanoscale Science, Engineering and Technology. US Government Printing Office. Washington, DC ©February 2011
b. Stanigner, Hildegarde. NANOTECHNOLOGY vs. ENVIRONMENTAL TECHNOLOGY World of Opportunities 2011 NREP/OIP Annual Conference and Workshop, Las Vegas, Nevada. ©October 5-6, 2011

21) Supplement to the President's FY 2012 Budget. The National Nanotechnology Initiative: Research and Development Leading to a Revolution in Technology and Industry. US Government Printing Office. Washington, DC ©February 22, 2011

22) Report of the National Nanotechnology Initiative Workshop (May 5-7, 2009). Nanotechnology-Enabled Sensing. National Nanotechnology Coordination Office. Arlington, VA ©2009

23) National Nanotechnology Coordination Office. Nanotechnology: Big Things from a Tiny World. Arlington, VA ©2008/2009

24) Hur, E. and DJ Moon. *Steam reforming of glycerol into hydrogen over nano-size Nibased hydrotalcite-like catalysts*. J. Nanosci. Nanotechnology 2011 Aug: 11(8): 7394-8 (http:///www.ncbi.nlm.nih.gov/pudmed/22103204)

25) Rivera, JA, Fetter, G and P. Bosch. *New hydroxyapatite-hydrotalcite composites II Microwave irradiation effect on structure and texture*. Journal of Porous Materials. Volume 16, Number 4, 409-418, DOI: 10, 1007/s10934-008-9213-z

26) Yin, Qing-Zhu; Cin-Ty, Aeolus Le; and Ulrich Ott. *Signatures of the s-process in presolar silicon carbide grains: Barium through Hafnium*. The Astrophysical Journal, 647:676-684, 2006 August 10
27) Dutta, Binay K., Tayseir M., Abd Ellateif, and Saikat Maitra. *Development of a porous silica film by sol-gel process*. World Academy of Science, Engineering and Technology 73 2011.

28) Dalla-Bona, Alexandra; Primbs, Jacqueline, and Angelina Angelova. *Nanoencapsulation of proteins via self-assembly with lipids and polymers*. Wiley-VCH Verlang GMBH & Co. KGaA, Weinheim, Germany © 2010 pgs 32-36

29) Dobrzhinetskaya, LF; Green, HW; Bozhilov, KN; Mitchell, TE and RM Dickerson.

Crystallization environment of Kazakhastan microdiamond: evidence from nanometric inclusion and mineral associations. Blackwell Publishing, Ltd. 0263-4929/03. Journal of Metamorphic Geology. Volume 21, Number 21, Nov 5, 2003

30) http://en.wikipedia.org/wiki/graphene, pgs 1 to 26, ©12/29/2011

31) Bostwirck, A. et al. *Symmetry breaking in few layer graphene films*." New Journal of Physics 9 (10) 385, ©2007

32) Zhou, S.Y., et. al. *First direct observation of Dirac fermions in graphite*. Nature Physics 2 (9): 595-599. ©2006

33) Sutter, P. *Epitaxial graphene: "How silicon leaves the scene*." Nature Materials 8 (3); 171 ©2009

34a) Morozov, S.V. et. al. Strong suppression of weak localization in graphene. Physical Review Letters 97 (1): 016801 ©2006

34 b) Kim, Kuen Soo; et al. "*Large-scale pattern growth of graphene films of stretchable transparent electrodes."* Nature 457 (7230) ©2009

35) Sutter, P. *Epitaxial graphene: "How silicon leaves the scene*." Nature Materials 8 (3); 171 ©2009

36) High-yield production of graphene by liquid-phase exfoliation of graphite: abstract:
NatureNanotechnology http://www.nature.com/nnano/journal/v3/n9/abs/nano.2008.215.html)

37) Brandenburg K. and U. Seydel. *Fourier Transform infrared spectroscopy of cell surface polysaccharides*. In: Infrared spectroscopy of biomolecules,(H.H. Mantsch and D. Chapman eds) 203-238; New York: Wiley-Liss ©1996

38) Bertoluzza A. Fagano C, Morellic MA, Gottadi, V. and MJ Guglielni. *Raman and infrared spectra on silica gel evolving toward glass*. J. Non-Cryst. Solids, 48, 117-128 ©1982

39) Perry CC. *Biogenic silica*. In: biomineralisation, Chemical and biological Perspectives. S. Mann, J. Webb and RJP Williams (eds). Chapter 8 pgs 233-256. VCH, Weinheim, Germany ©1989

40) Panick G. and R. Winter. *Pressure-Induced Unfolding/Refolding of Ribonuclease A: Static and Kinetic Fourier Transform Infrared spectroscopy Study*. Biochemistry, 39/7, 1862-1869 ©2000

41) Padmaja, P. Anikumar GM, Mukundan P, Aruldhas G and KGK Warrier. *Characterization of stoichiometric sol-gel mullite by Fourier transform infrared spectroscopy*. Int. J. of Inorg Materials 3, 693-698 ©2001

42) www.springerlink.com/index/G2266H767760104V.pdf Polyethylene-polystyrene gradient polymers II The swelling. ~730 cm-1 band
www.ieexplore.ieee.org/iel5/84/32479/01516171.pdf?arnumber=1516171
Polymerbased wide-bandwidth and high-sensitivity micro machined tunneling

accelerometer, chemical sensor, infrared (IR) radiation sensor.

43) www.ieeexplore.ieee.org/iel5/3/29490/01337030.pdf?arnumber=1337030. Mid-IR Optical Limiter based on Type II Quantum Wlls ~ 1493 frequency. www.ammin.geoscienceworld.org/content/96/11-12/1856.full Methods to analyze metastable and microparticulate hydrated related to water ~ 1463 frequency.

45) www.minsocam.org/msa/ammin/toc/.../Hammer_p569-576_98.pdf Single-crystal IR spectroscopy of very strong hydrogen bonds in pectolite at ~ 1396 cm-1 frequency.

46) www.ieeexplore.ieee.org/xpls/abs_all.jsp?arnumber=5441498 Bragg mirrors consisting of five altering layers of Ir and porous SiO2 are designed. ~ 1535 - 1538 frequency.

47) www.a1securitycameras.com/Samsung-Techwin-SND-3080.html www.geo.uscb.edu/facutly/awramik/poubs/!G!SO611.pdf ~3080 cm-1 bandwidths of carbonaceous material varied according to the degree of its in situ characterization.

49) Benning, Liane G.; Phoenix, VR; Yee n. and MJ Tobin. *Molecular characterization of cyanobacterial silicificaiton using sychrontorn infrared microscopy.* Geochimica Cosmochimica Aeta. ©April 16, 2003 Sychorotron-based FTIR of cyanobacteria.

50) Kneipp, Janina, et. al. *In situ indentification of protein structural changes in prioninfected tissue*. Elsevier B.V. Biochimica et Biophysics Aeta 1639 (2003) 152-158

51) http://www.cuttingedge.org/News/n1875.cfm Title: DARPA is Funding an Implantable Chip Far more Advanced than "Digital Angel" MMEA Multiple micro electrode array is far advanced it can fulfill Rev 13:16-18? Part 1 of 5 parts. ©12/29/2011

52) www.mtixtl.com Processing Routes for Aluminum based Nan-Composites. EXTEC green phenolic materials, 1100 degrees Celcius ©2010 C. Borgonova

53) Kim, Jang Sub. Samsung Electronics, Co. LED US Patent: 2007/0180954 A1 published 09-Aug-2007. Copper nano-particles, method of preparing the same, and method of forming copper coating film using the same.

54) Stratekis, E.; Ranelia A and C. Fotakis. *Biomimetric micro/nanostructured functional surfaces for microfluidic and tissue engineering applications (3D bioimetic modification materials).* American Institute of Physics ©30 March 2011 AIP Biomicrofluidics in Tissue Engineering and Regenerative Medicine.

55) Graham, S. et. al. *An enzyme-detergent method for effective prion decontamination of surgical steel*. Journal of General Virology. March 2005 vol. 86 no. 3 pgs 869-878

Figure 1 - An example of a mesogen in its architectural design concept phase. (FIGURE 1.1 taken from Vibrational Spectroscopy of Biological and Polymeric Materials by Vasilis G. Gregoriou and Mark S. Braiman, page 11. CRC Taylor & Francis. Boca Raton, FL ©2006)

Studying the Viscoelastic Behavior of Liquid Crystalline Polyurethanes

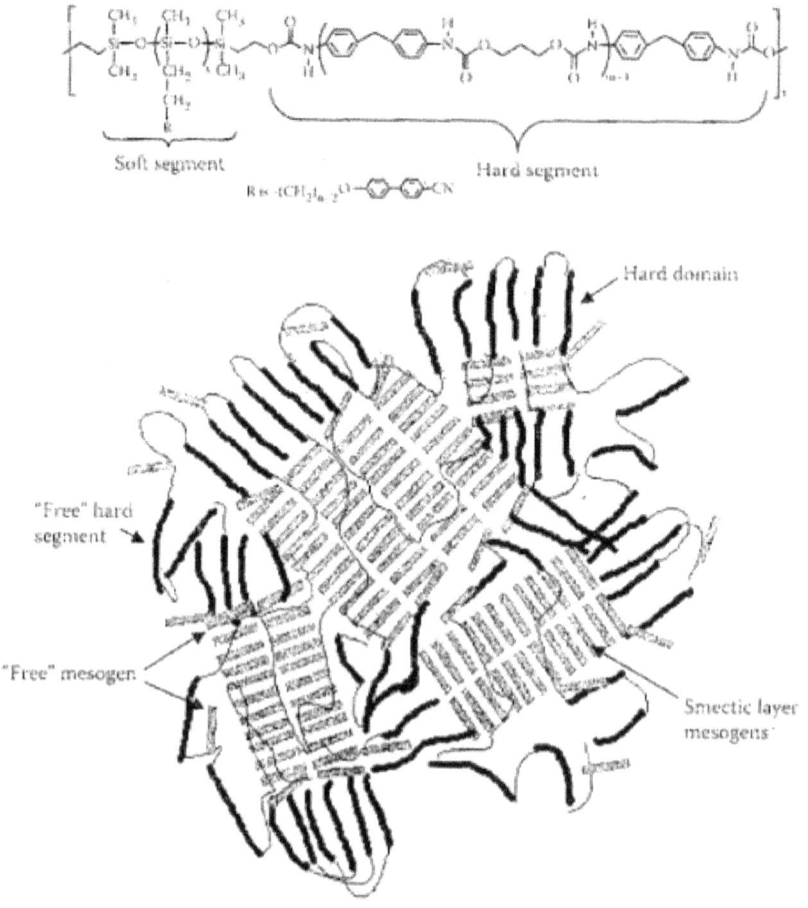

FIGURE 1.1 (a) Chemical structure of side chain liquid crystalline polyurethane. (b) Polyurethane morphology representation. (Reproduced Nair, B. R.; Gregoriou, V. G.; Hammond, P. T. *J. Phys. Chem. B.* **2000**, with permission. Copyright [2000], American Chemical Society.)

FIGURE 2 - **Acrylate side chain Mesogen**

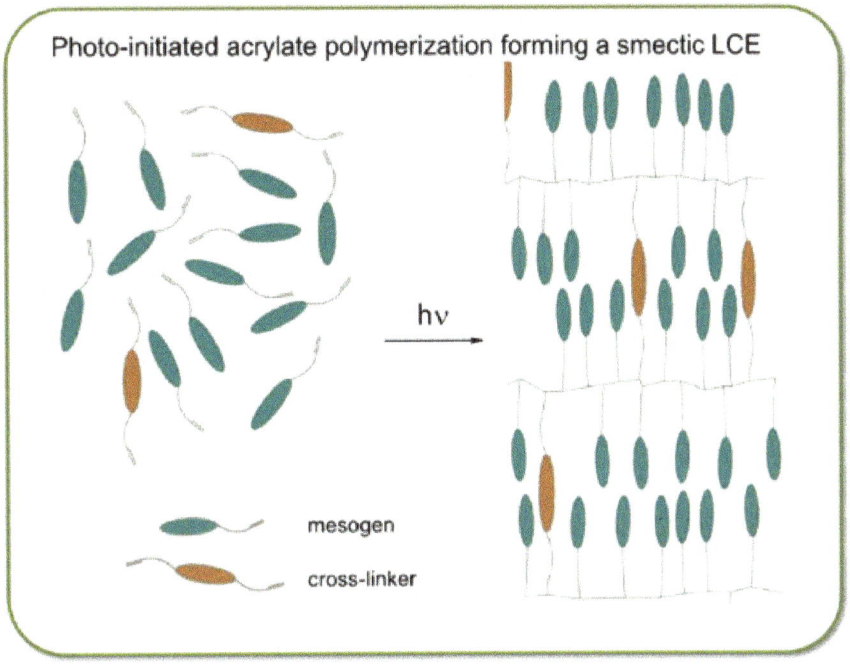

FIGURE 3 - **Siloxane side chain Mesogen**

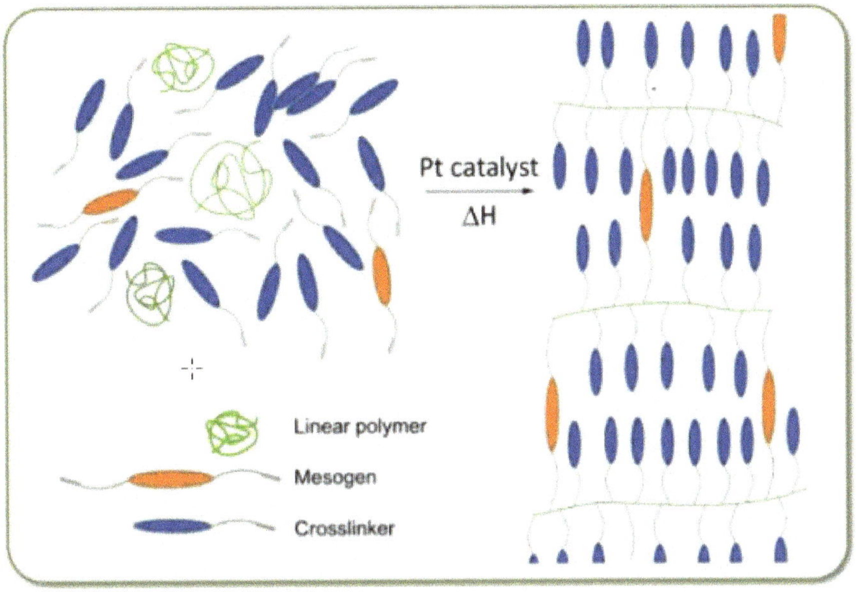

FIGURE 4 - Main chain Liquid Crystal forming Mesogen

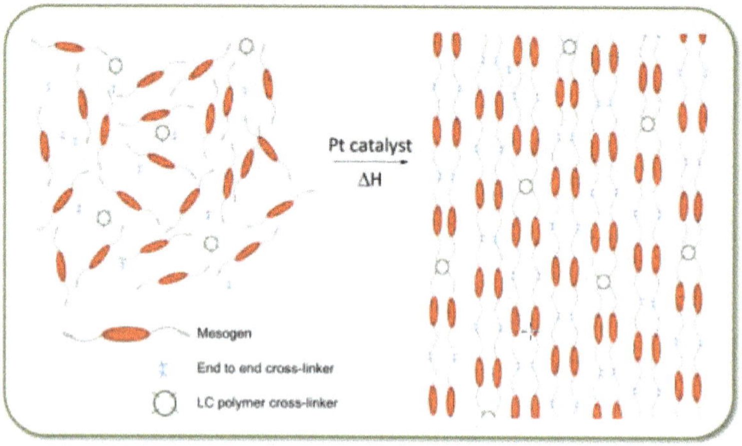

FIGURE 5 - Polyester/polyurethane structural reactions with alcohol

FIGURE 6 A & B - Smetic Phase of Nematic Mesogens

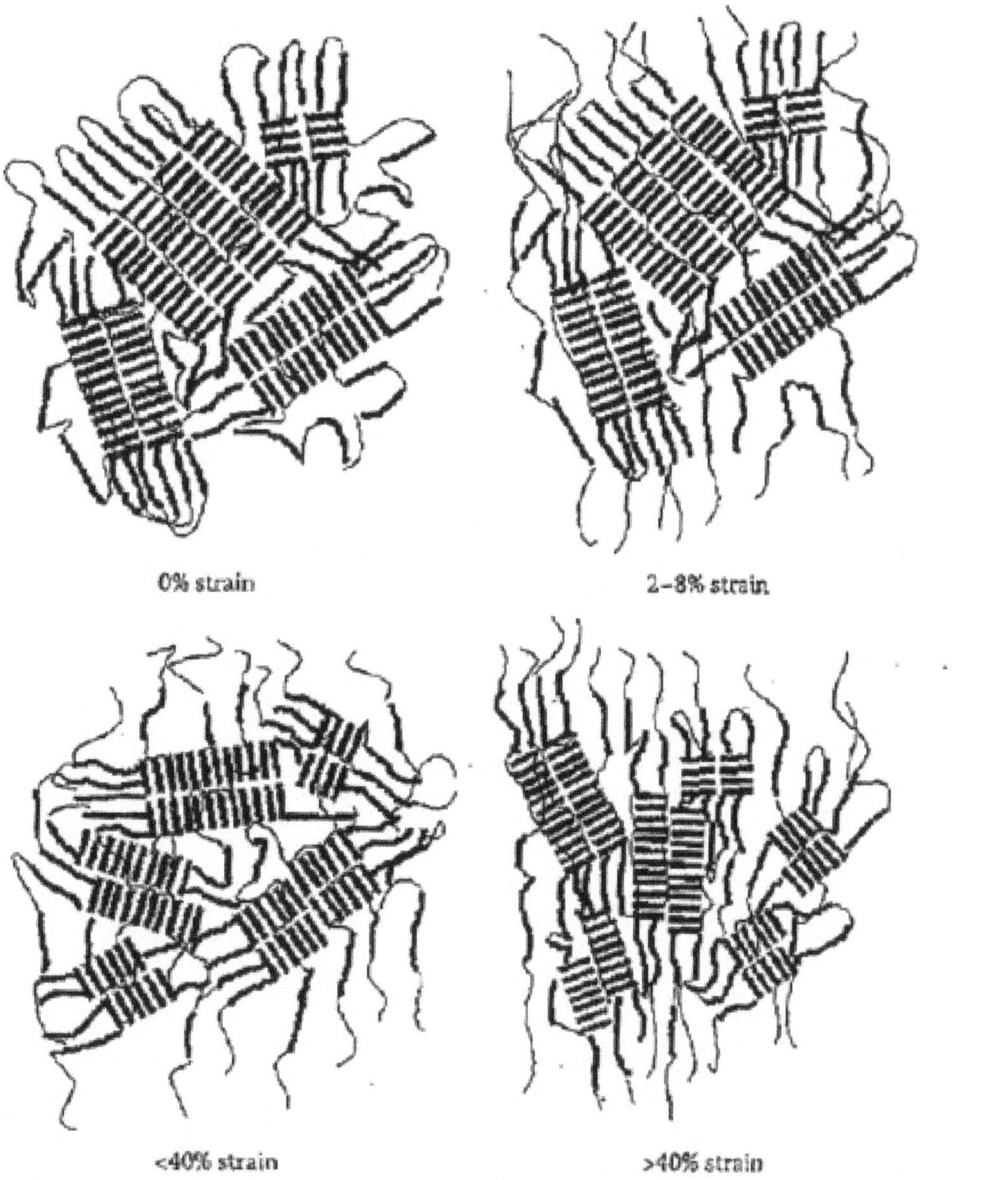

FIGURE 1.18 Proposed model of cooperative deformation of hard segments and smectic layers as a function of strain. (Reproduced from Nair, B. R.; Gregoriou, V. G.; Hammond, P. T. *Polymer* 2000 with permission. Copyright [2000] Elsevier.)

(FIGURE 1.18 taken from Vibrational Spectroscopy of Biological and Polymeric Materials by Vasilis G. Gregoriou and Mark S. Braiman, Page 31. CRC Taylor & Francis. Boca Raton, FL ©2006)

FIGURE 7 - Chiral Phase of Nematic Phase of Mesogens

FIGURE 8 - Other Image of Chiral Phase of Nematic Phase of Mesogens
(Note pitch distance)

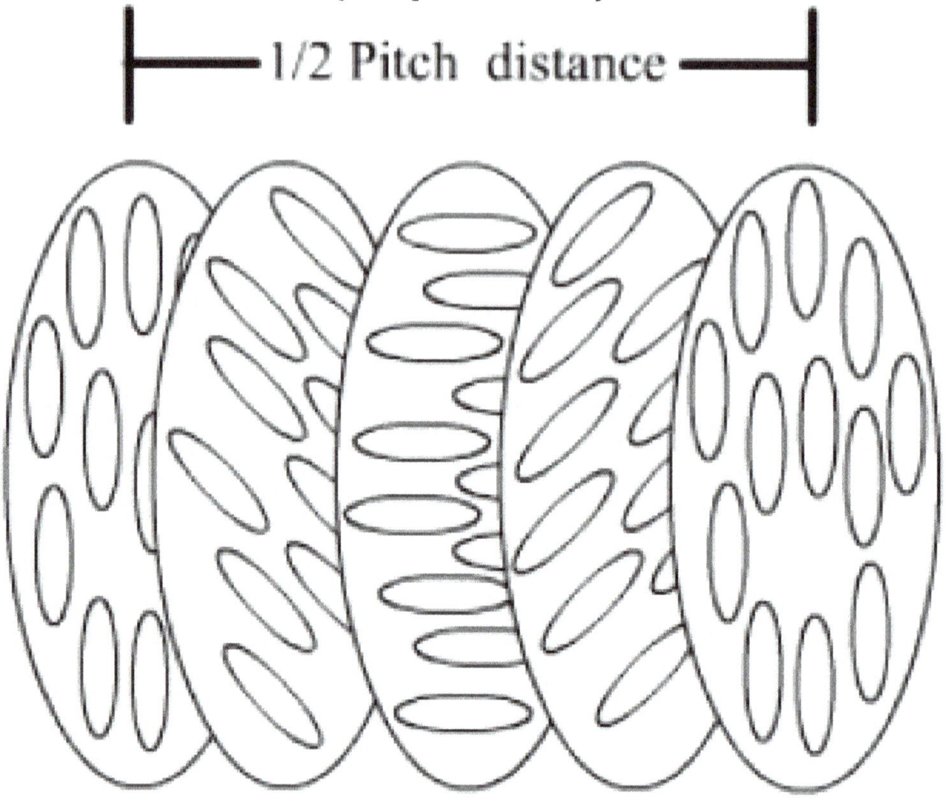

FIGURE 9 - TEM images of silica "bundled" arrays and cages. Taken from the article Silica Nanotubes and Nanofiber Arrays, by L. Zhong Wang, *et al*, Advanced Materials 200, 12, No. 24, December 15.

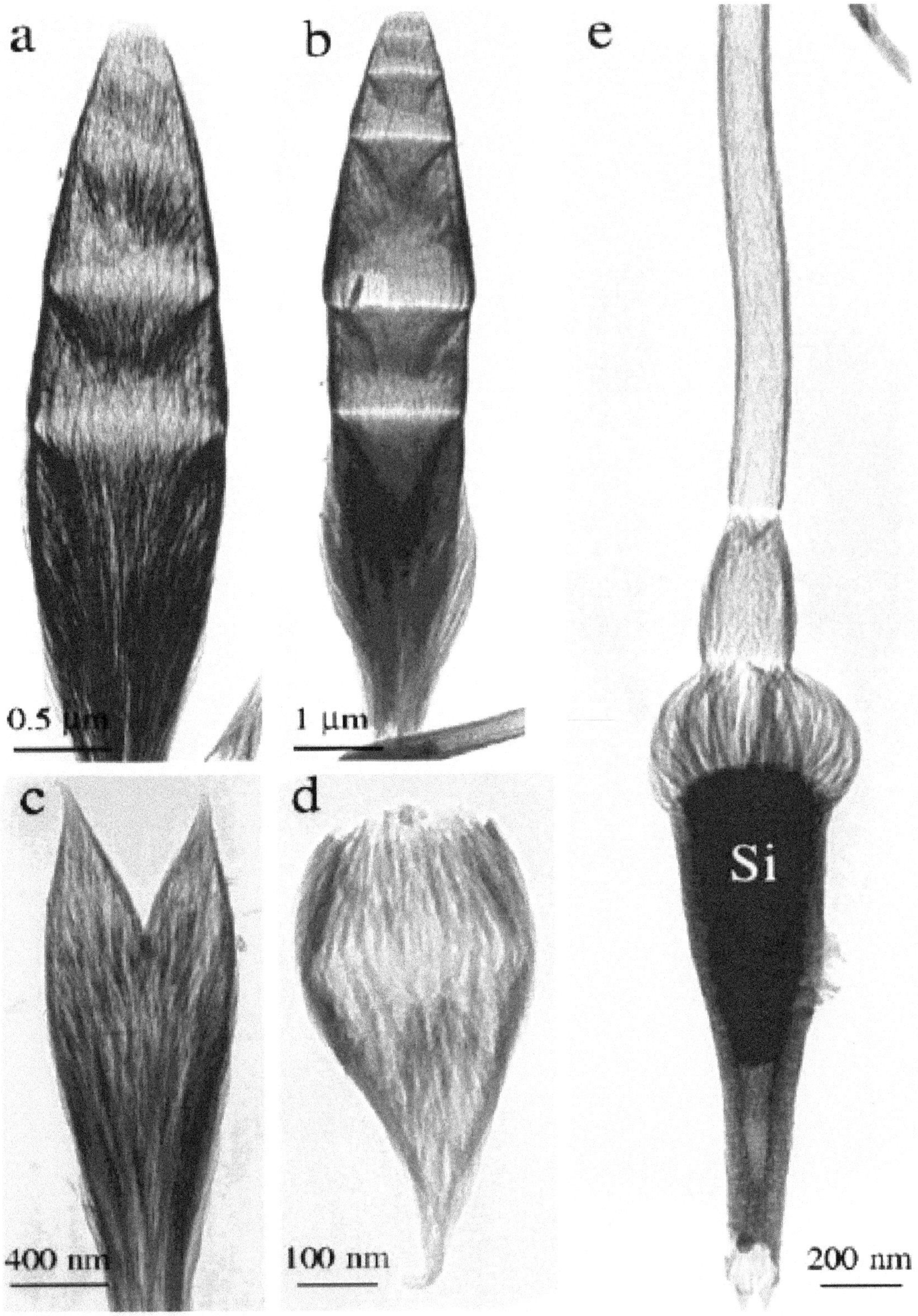

TEM images of synthesized silica nano-tube structures usually formed following the trapping of silicon nano-crystals

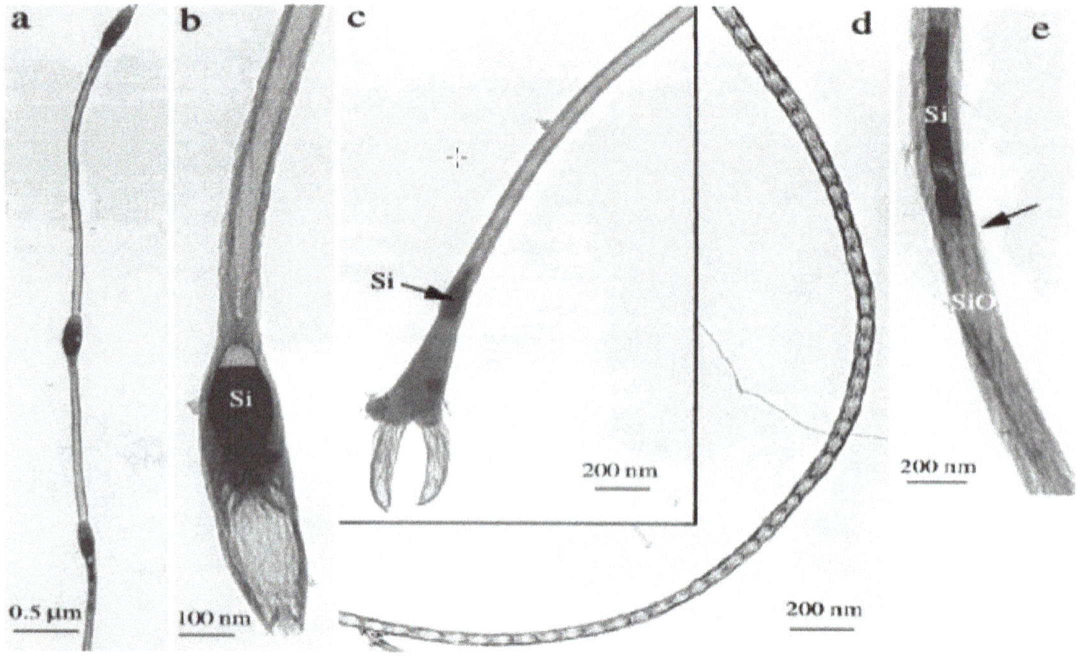

FIGURE 10 - Silica, Silicone and Silicon photomicrographs of "Chinese Lanterns" as taken from skin biopsy specimens. Photomicrograph taken from Project Fiber, Morgellons and Meteorite, ©2006 by Dr. Hildegarde Staninger TM, Phase I of Project.

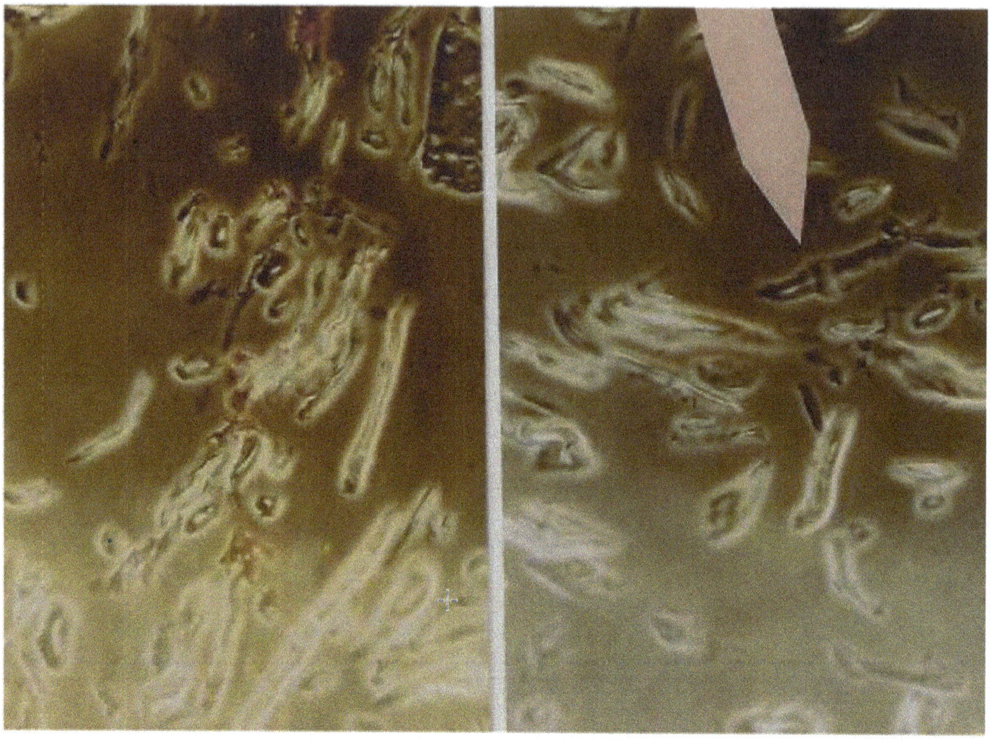

FIGURE 11 - **RAMAN/Micro FTIR Chromatograph of nano hooks/claws of nasal bulb mesogens collected from Los Angeles, CA; NYC, New York; and Saudi Arabia in June 2010 for advanced materials testing.** Analysis and photomicrographs by *Applied Consumer Services, Inc.*, Hialeah Gardens, FL, for *Integrative Health Systems TM, LLC*, Los Angeles, CA.

Los Angeles Specimen

17733

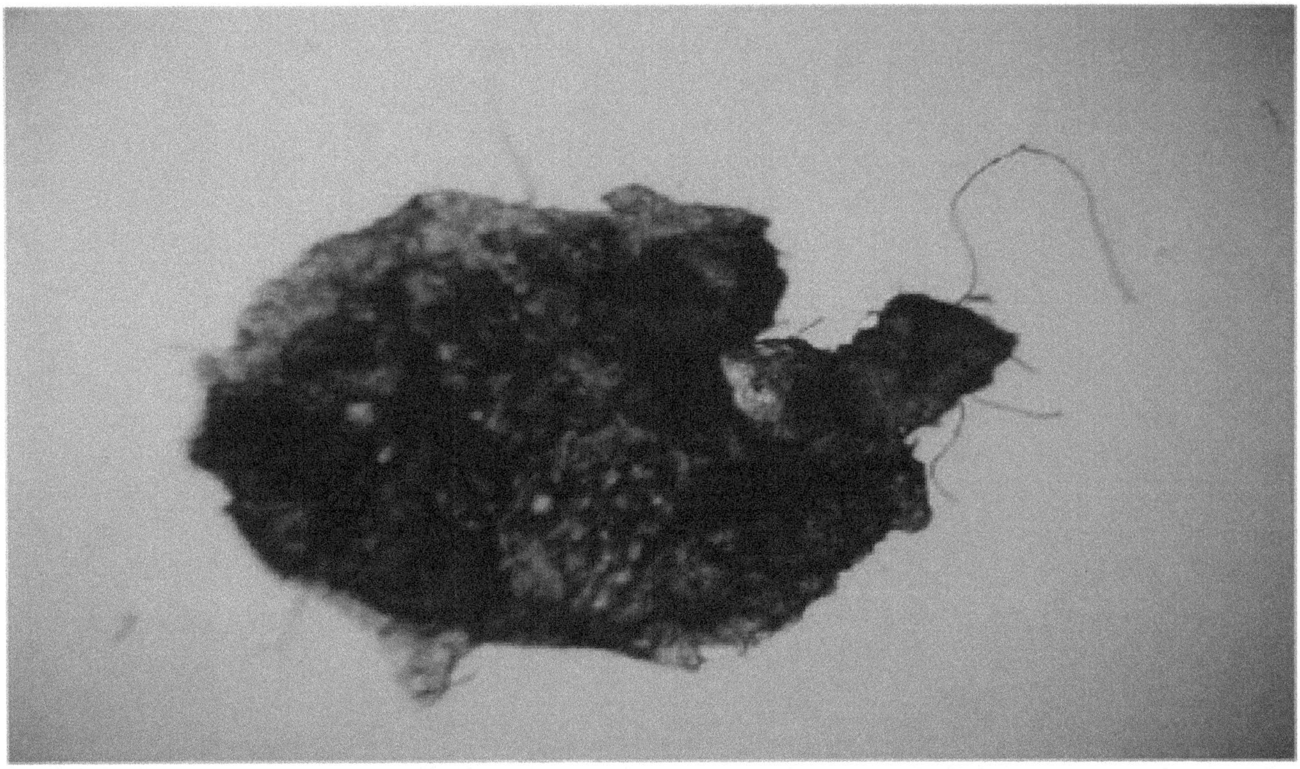

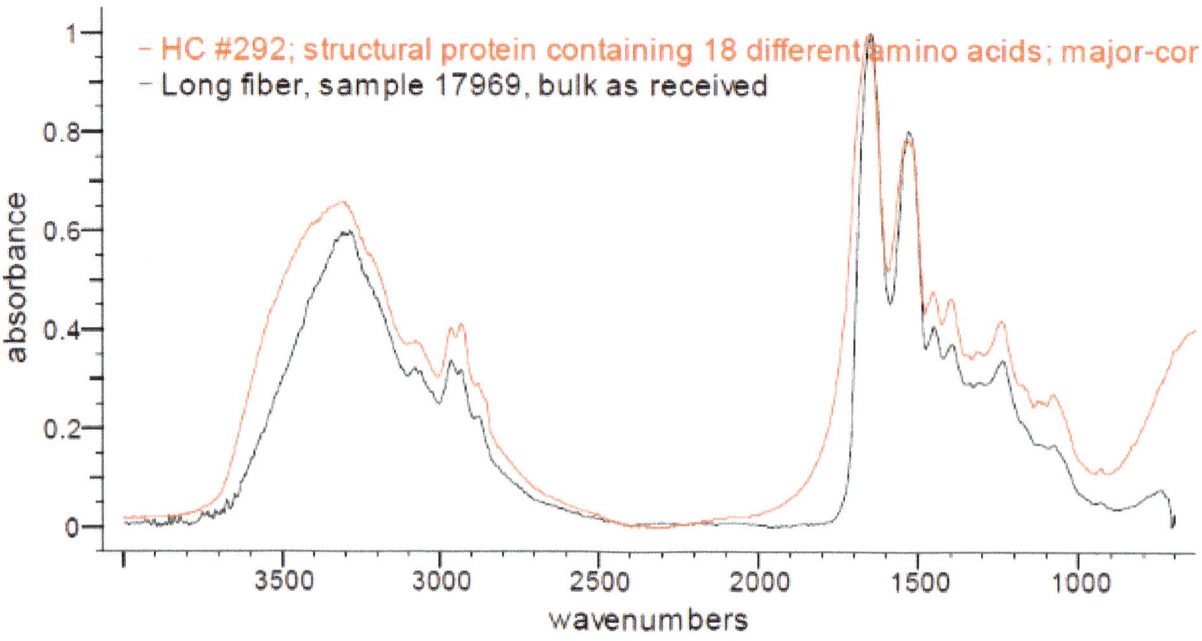

Specific assignments for protein bands

3300 cm-1	NH stretch
3080 cm-1	CNH overtone
3000 – 2850 cm-1	CH stretch
1650 cm-1	C=O
1535 cm-1	CNH
1396 cm-1	CH rock
1493 cm-1	CH3
1240 cm-1	CNH
730 cm-1	NH wag

FIGURE 12 - Micro FTIR Analysis result of Brain Menginoma (Mesogen Biosensor). Analysis performed by *Applied Consumer Services, Inc.*, Hialeah Gardens, FL ©2011.
December 6, 2011

L/N 19666

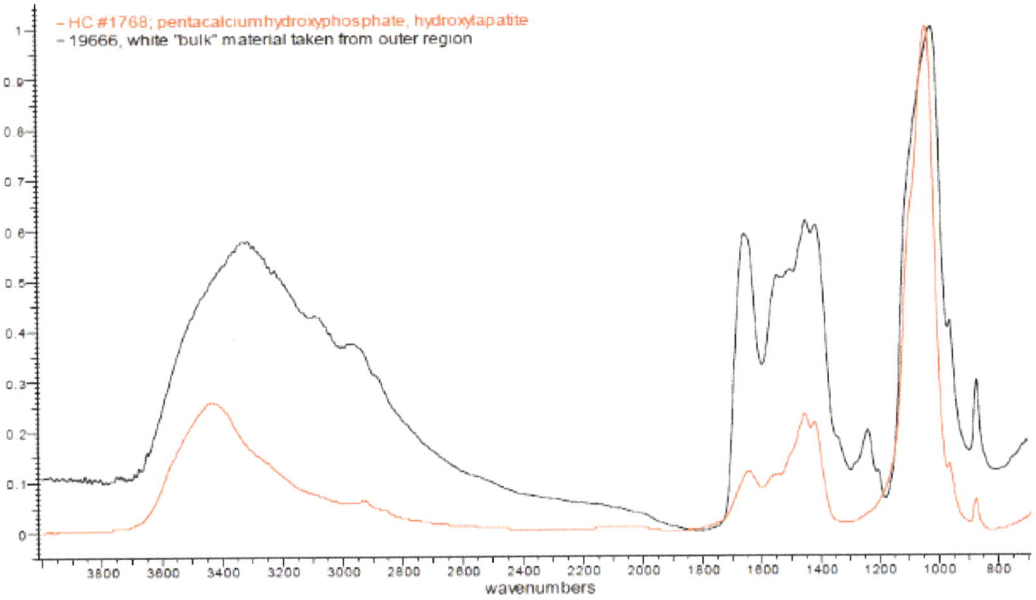

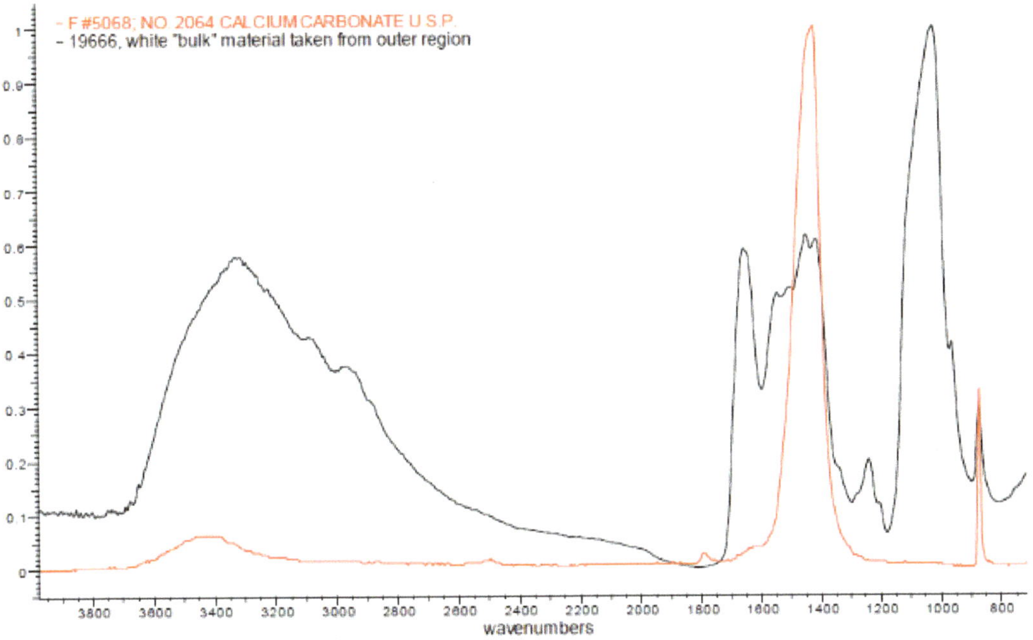

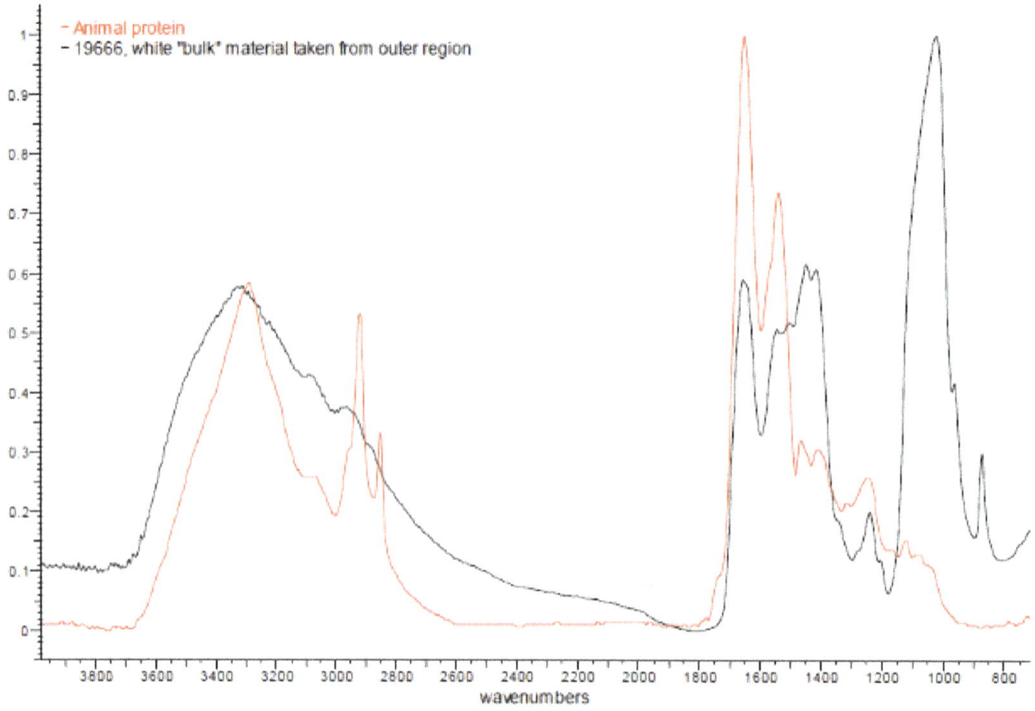

Animal protein
19666, white "bulk" material taken from outer region
1
0.9
0.8
0.7
0.6
0.5
0.4
0.3
0.2
0.1
0
3800
3600
3400
3200
3000
2800
2600
2400
2200
2000
1800
1600
1400
1200
1000
800
wavenumbers

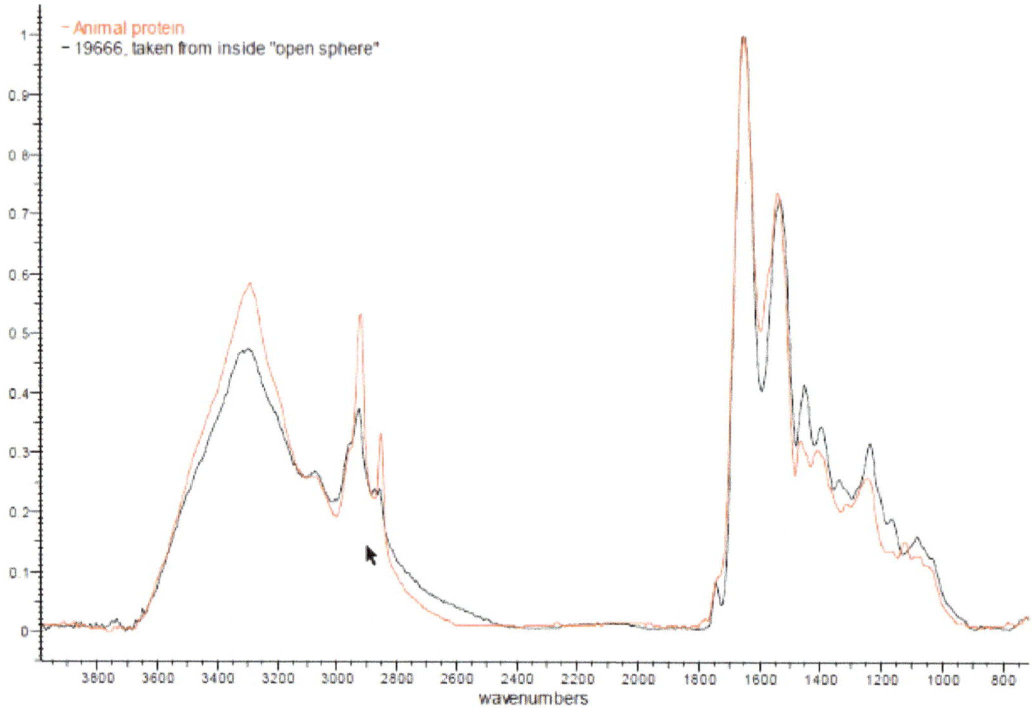

Animal protein
19666, taken from inside "open sphere"
1
0.9
0.8
0.7
0.6
0.5
0.4
0.3
0.2
0.1
0
3800
3600
3400
3200
3000
2800
2600
2400
2200
2000
1800
1600
1400
1200
1000
800
wavenumbers

RESULTS AND OBSERVATIONS

□ The outside layered skeleton - sample 19666 - was highly consistent with a hydroxyapatite (pentacalcium hydroxyphosphate) with calcium carbonate and animal protein. This mixture suggests that it originated from a deposit such as a kidney stone. Calcium phosphate in the hydroxyapatite form compose bone, teeth, and these calculi deposits.

□ The inside "open sphere" portion of sample 19666 was more of red color and consistent with animal protein.

FIGURE 13 - EDS/SEM results of Brain Menginoma – Brain Chip Mesogen Bio Sensor. Analysis performed by *Applied Consumer Services, Inc.*, Hialeah Gardens, FL ©2011

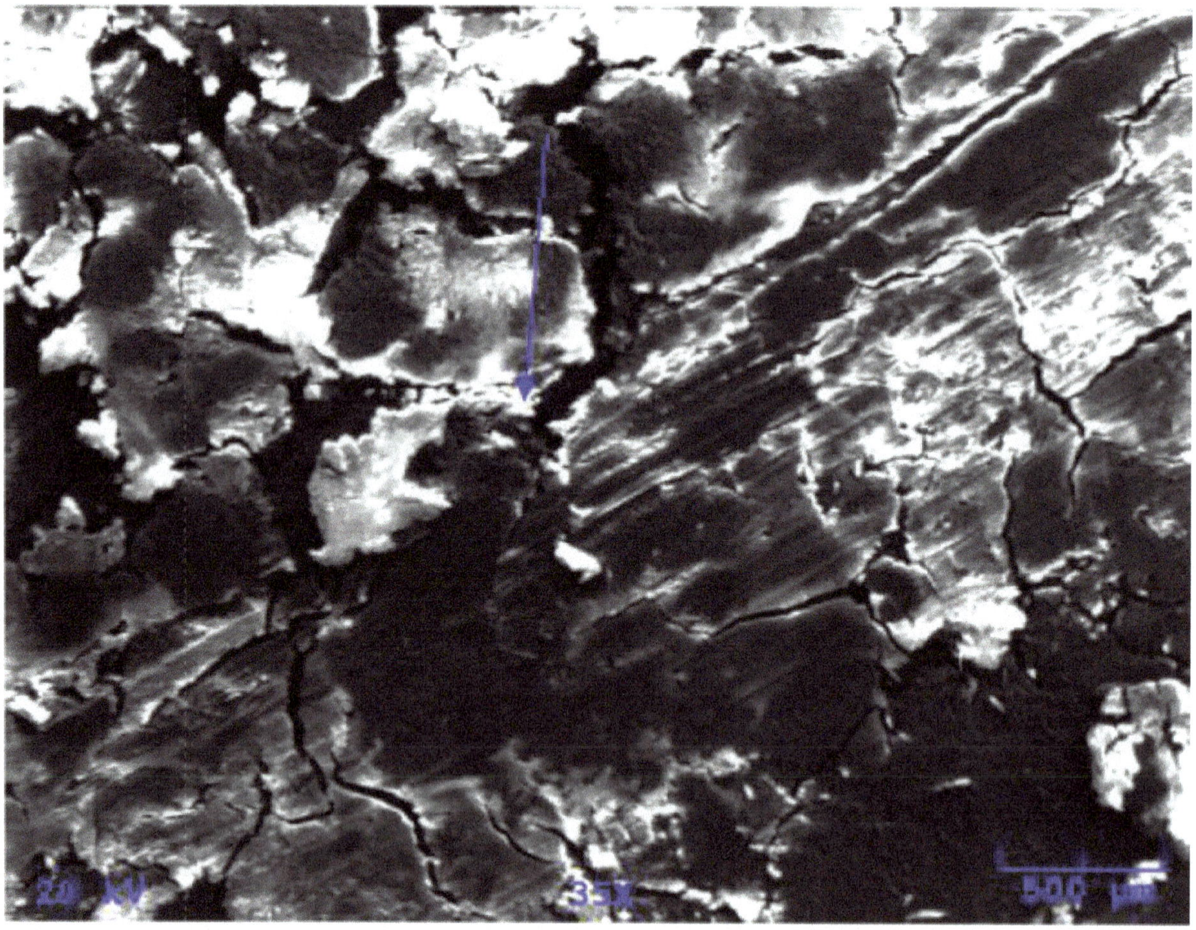

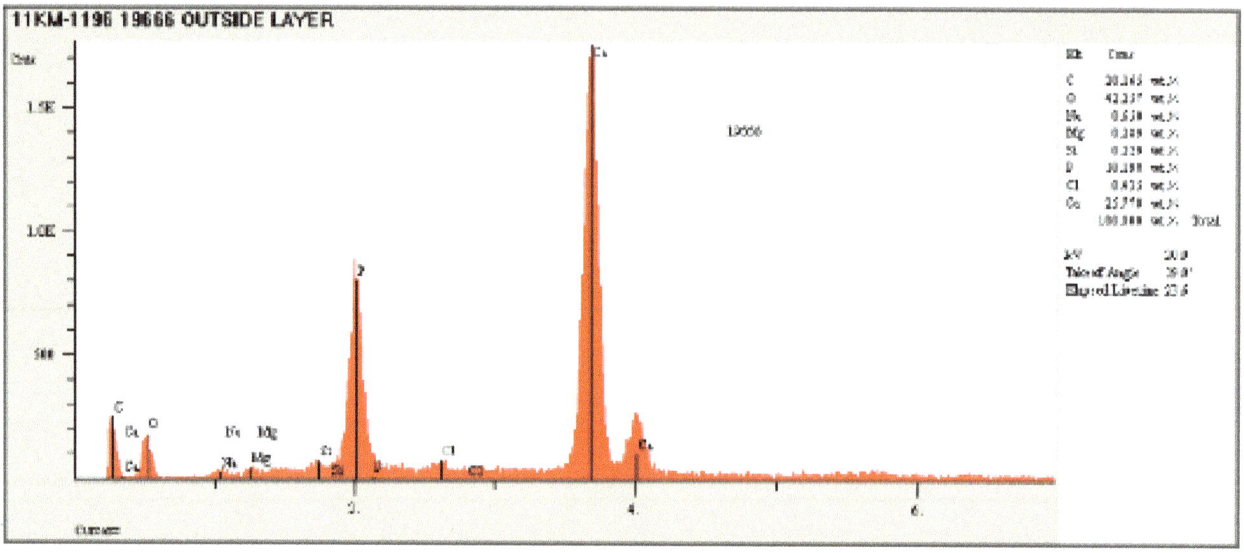

FIGURE 14 - EDS/SEM Results of Ear Canal Specimen. Analysis performed by Applied *Consumer Services, Inc.*, Hialeah Gardens, FL ©2011

Sample A

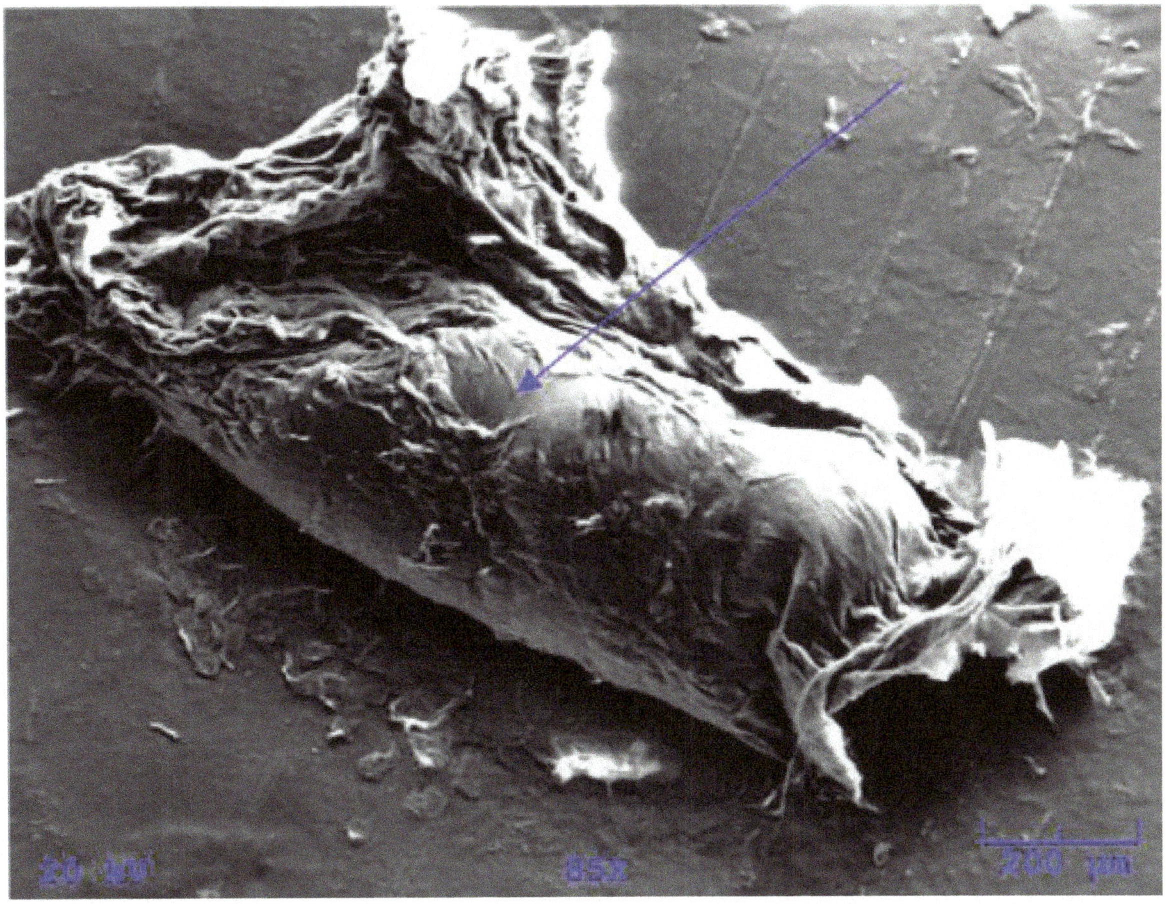

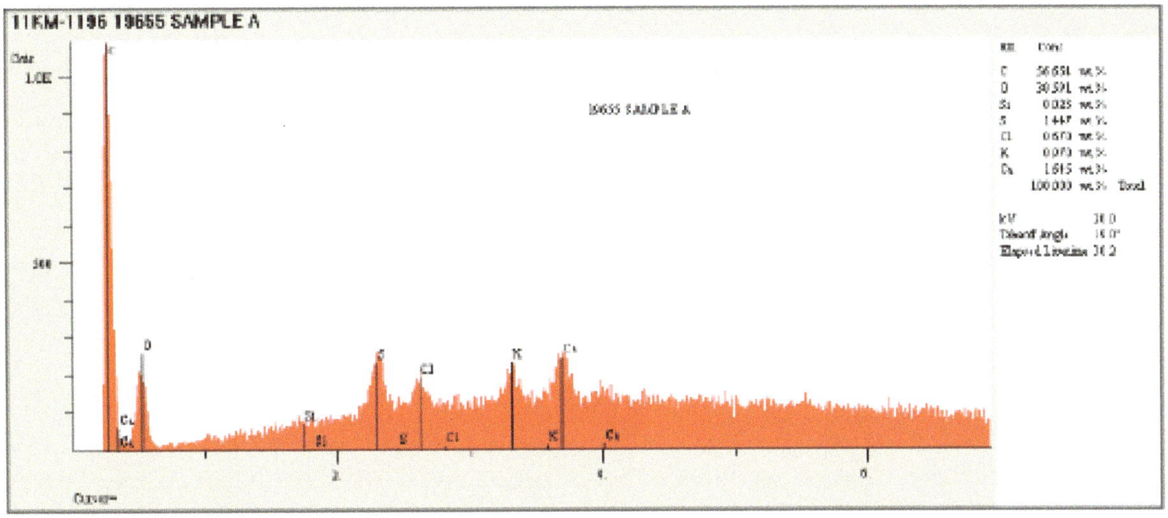

Sample B

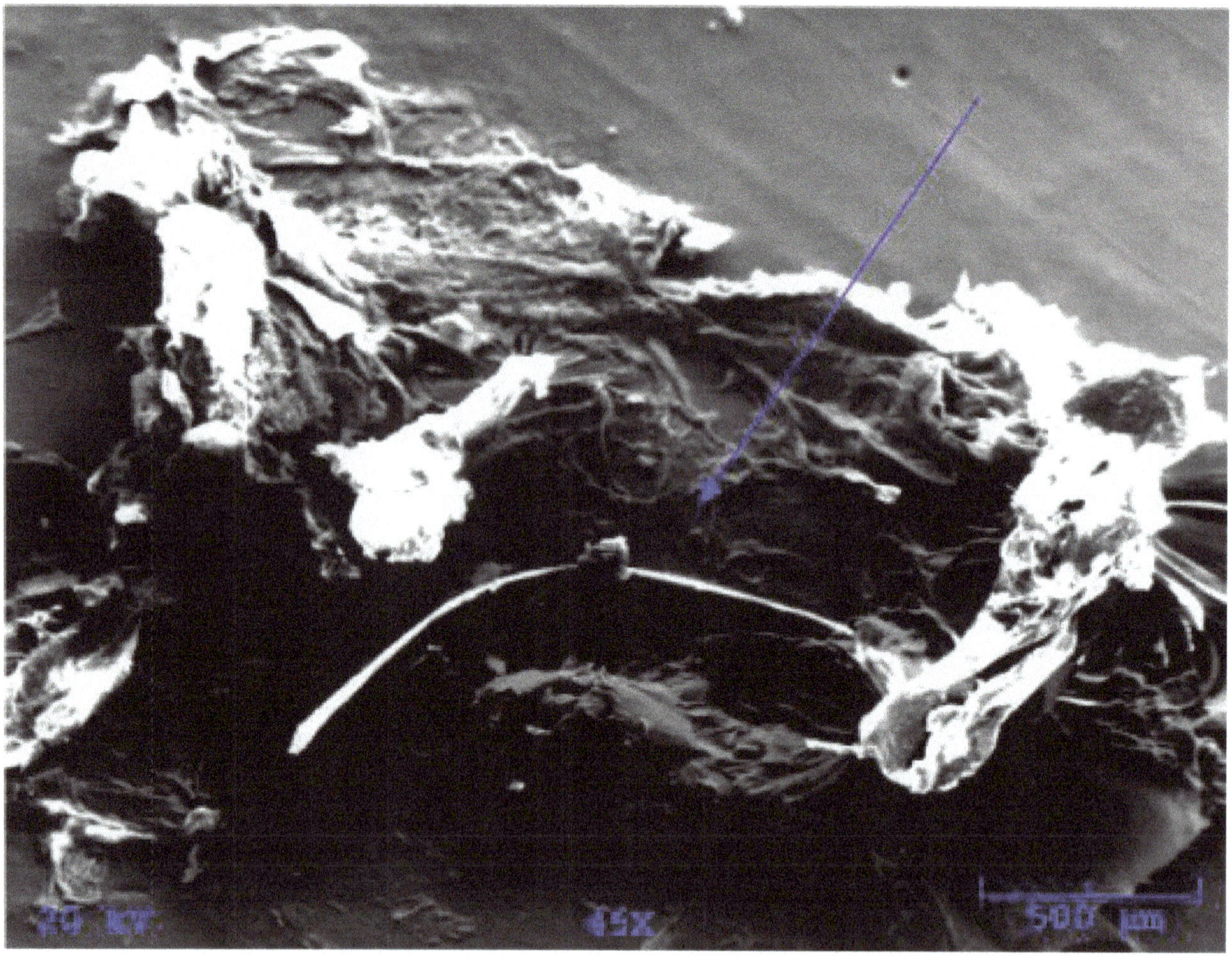

20 kV
45X
500 µm

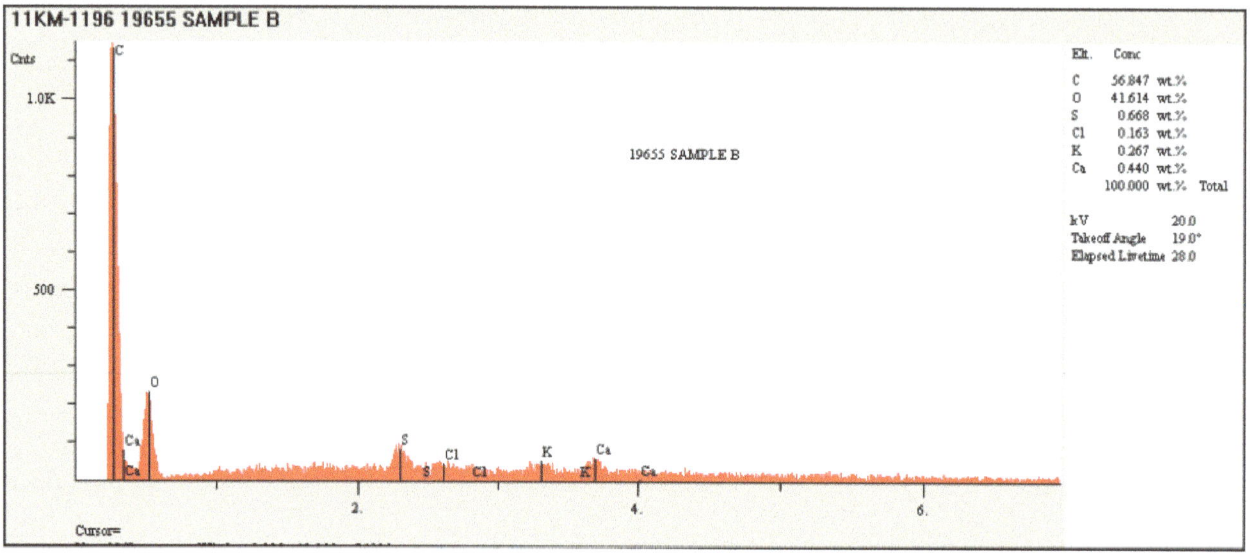

11KM-1196 19655 SAMPLE B
Cnts
1.0K
500
C
O
Ca
S
Cl
K
Ca
2.
4.
6.
Cursor=
19655 SAMPLE B
Elt. Conc
C 56.847 wt.%
O 41.614 wt.%
S 0.668 wt.%
Cl 0.163 wt.%
K 0.267 wt.%
Ca 0.440 wt.%
100.000 wt.% Total
kV 20.0
Takeoff Angle 19.0°
Elapsed Livetime 28.0

Sample B Fiber

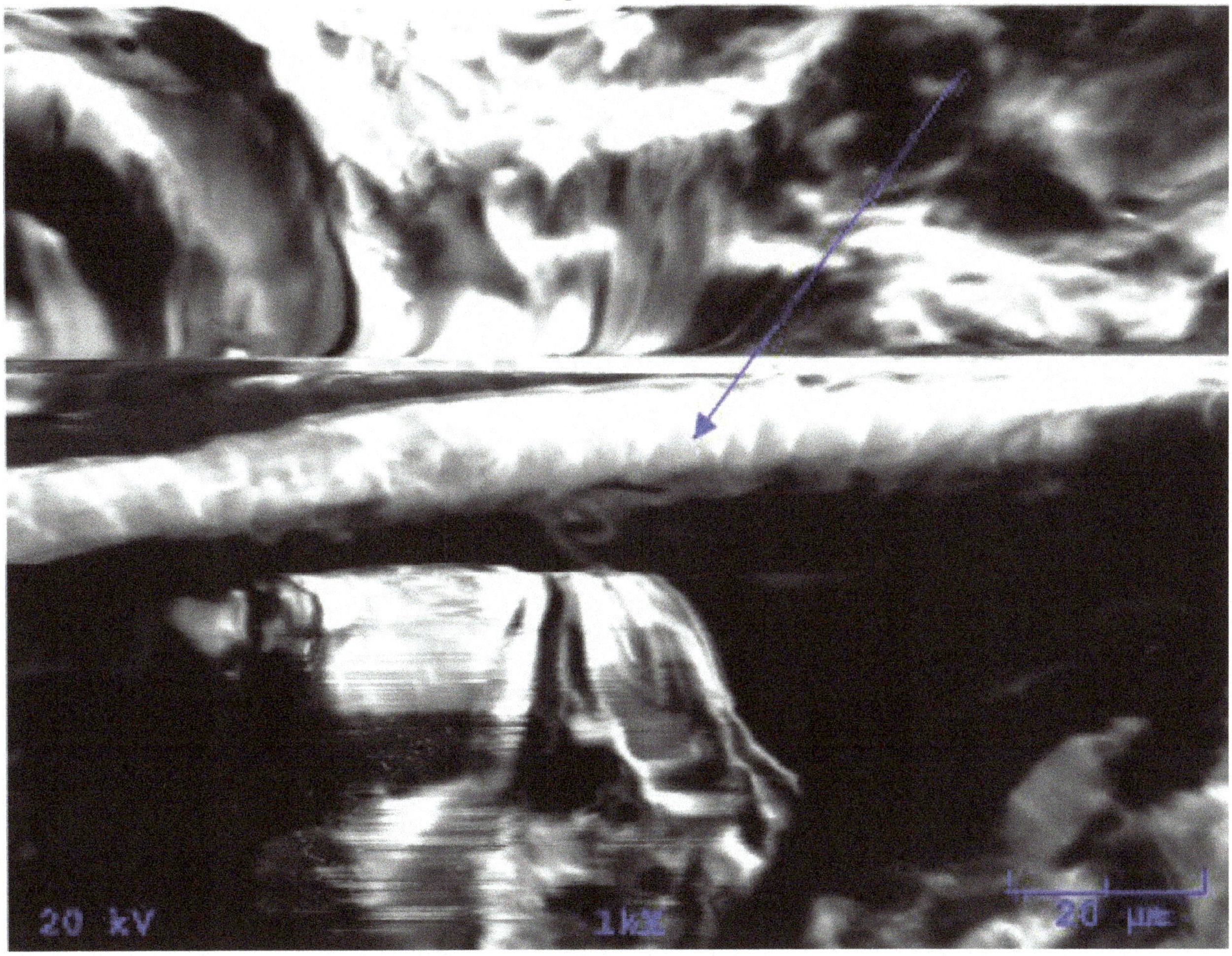
20 kV
1kX
20 µm

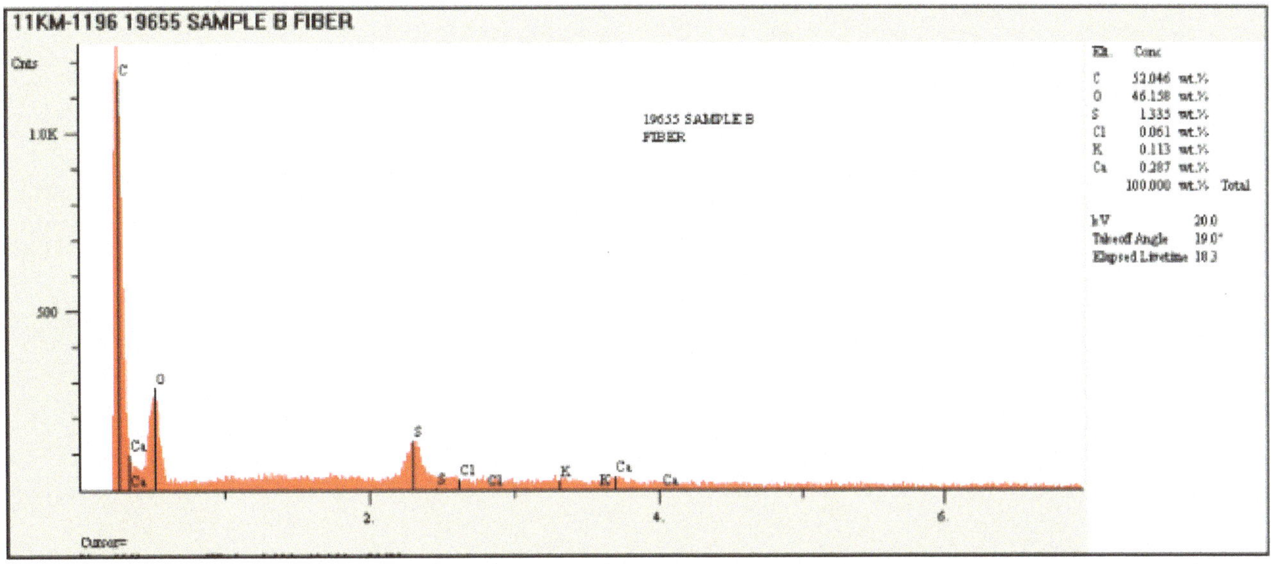
11KM-1196 19655 SAMPLE B FIBER
Cnts
1.0K
500
19655 SAMPLE B FIBER
C
O
Ca
S
Cl
K
2.
4.
6.
Cursor=
Elt. Conc
C 52.046 wt.%
O 46.158 wt.%
S 1.335 wt.%
Cl 0.061 wt.%
K 0.113 wt.%
Ca 0.287 wt.%
100.000 wt.% Total
kV 20.0
Takeoff Angle 19.0°
Elapsed Livetime 18.3

Wax Sample

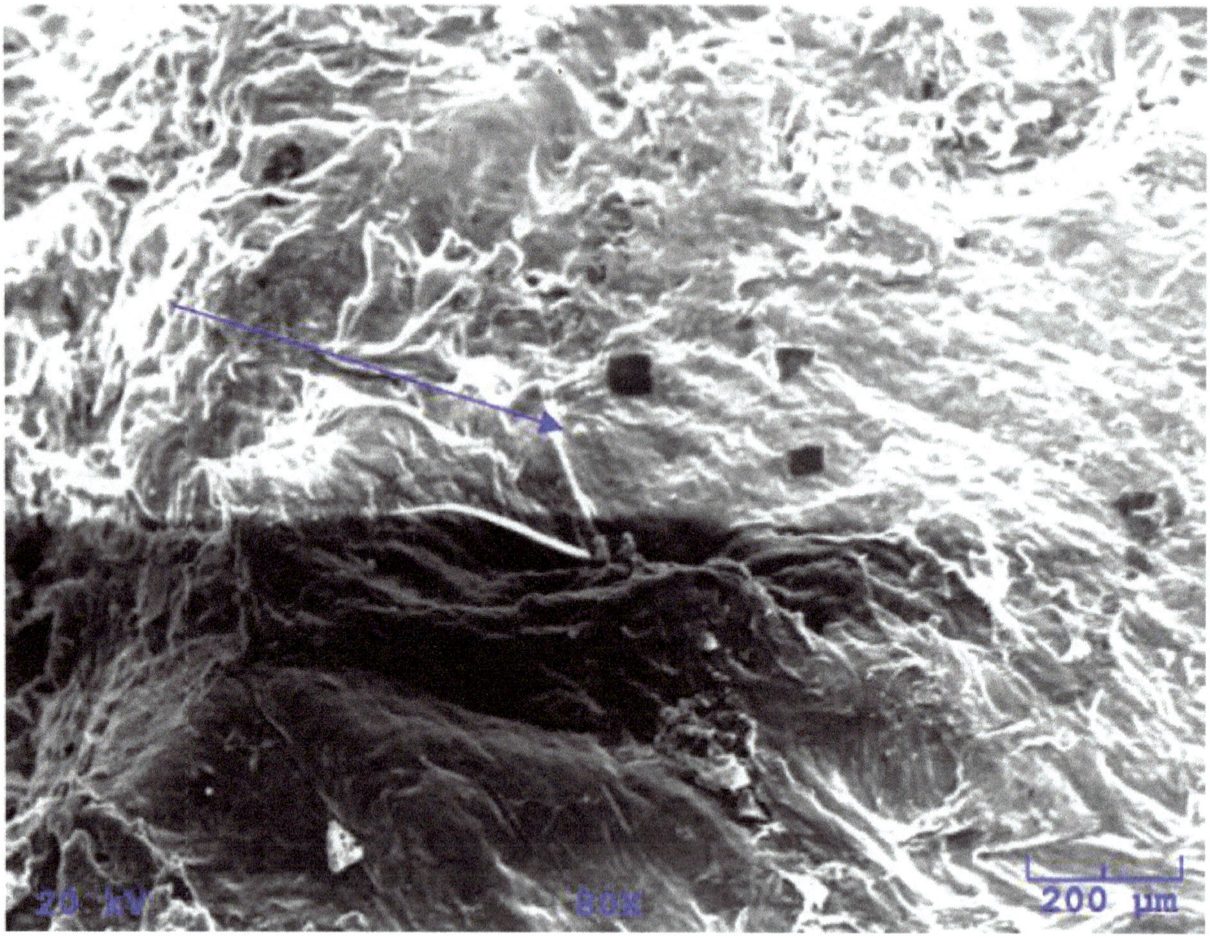

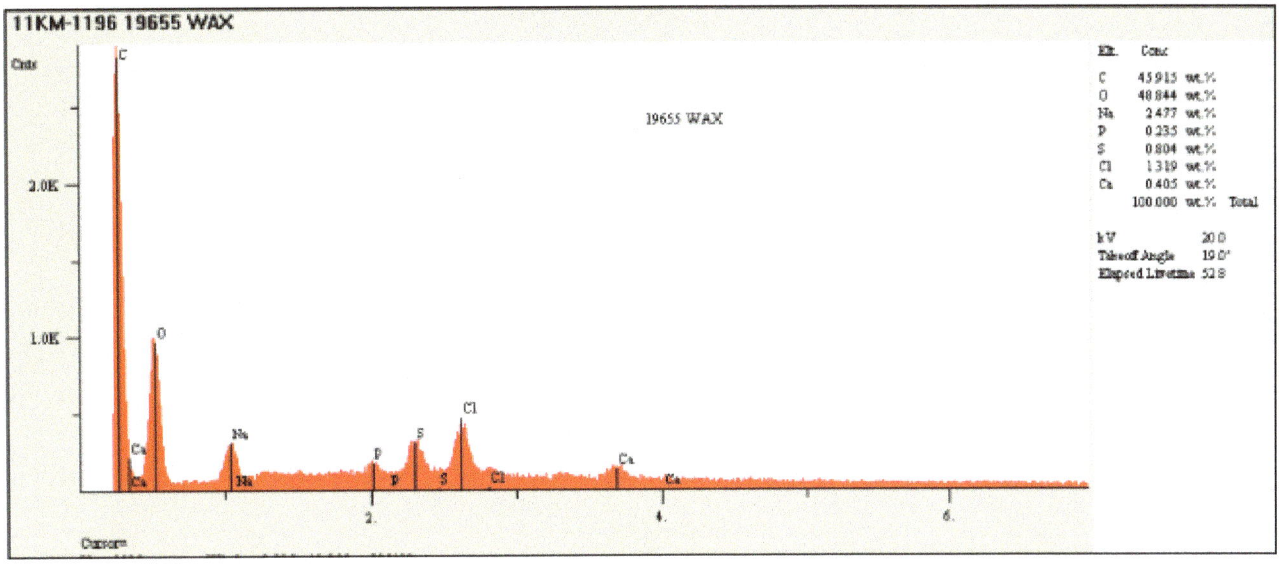

FIGURE 16 - RAMAN/MICRO FTIR Analysis of black square and other compounds from Ear Canal Specimen as a mesogen. Analysis performed by *Applied Consumer Services, Inc.*, Hialeah Gardens, FL ©2011

December 2, 2011 L/N: 19655/1B

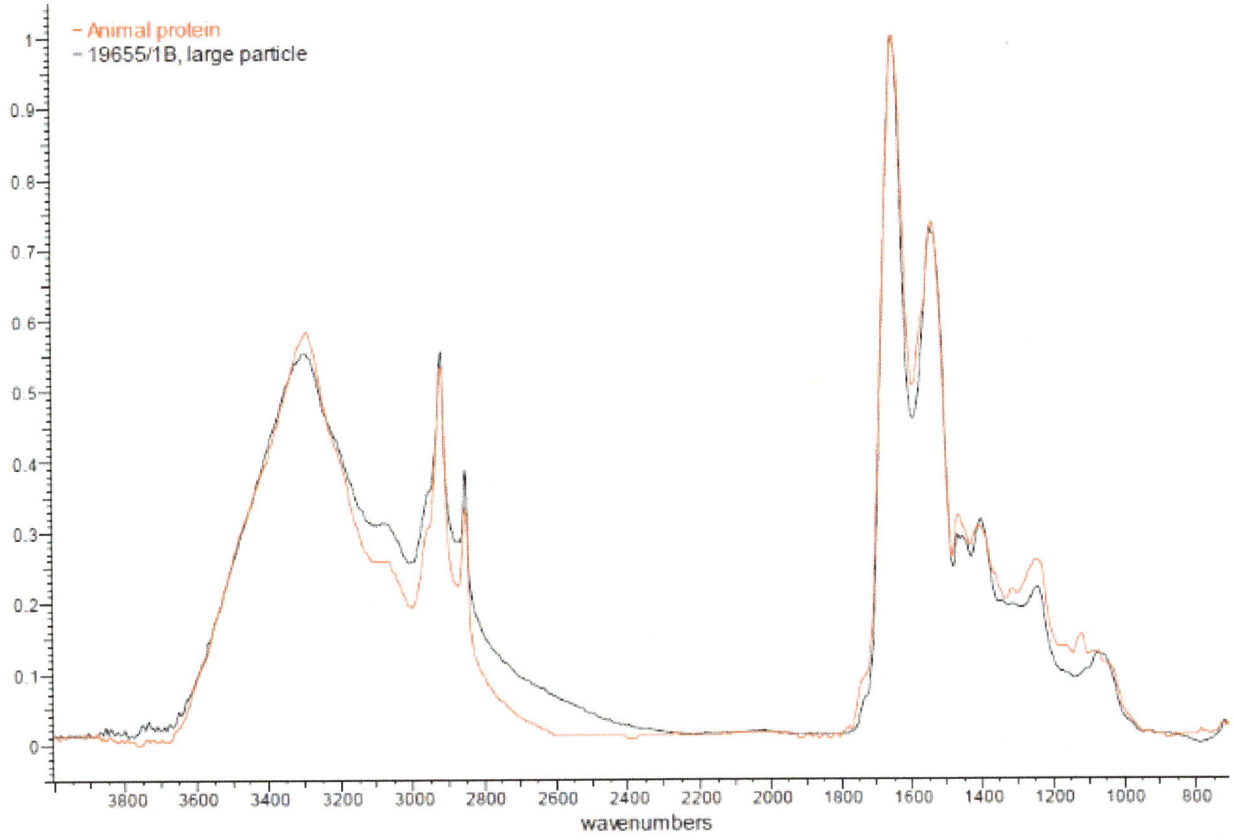

RESULTS AND OBSERVATIONS

Sample 19655/1B was highly consistent with an animal protein. Specific assignments for protein bands are:

3300 cm-1	NH stretch
3080 cm-1	CNH overtone
3000 – 2850 cm-1	CH stretch
1650 cm-1	C=O
1535 cm-1	CNH
1396 cm-1	CH rock
1493 cm-1	CH3
1240 cm-1	CNH
730 cm-1	NH wag

December 2, 2011 L/N 19655/2A and 19655/2B

FTIR spectrum – featureless except small hydroxyl OH

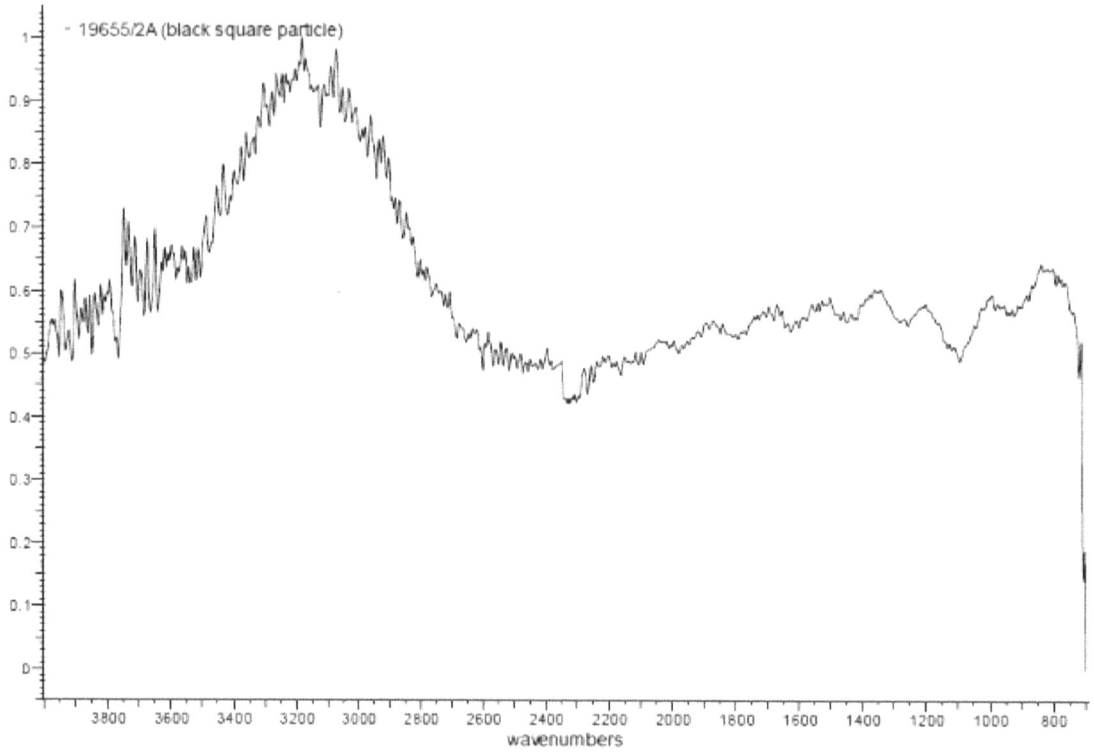

Raman spectrum – comparison to graphite

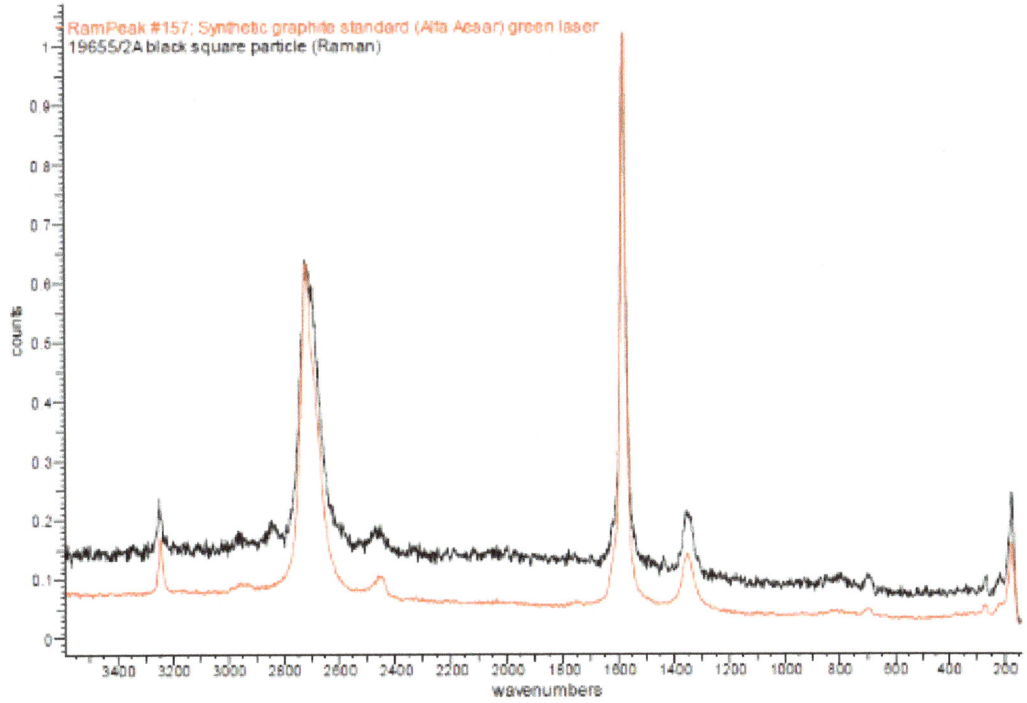

FIGURE 17 - Moon Glycerine U.S.P. Ear Canal Specimen. Micro FTIR (BioRad). Analysis performed by *Applied Consumer Services, Inc.*, Hialeah Gardens, FL ©2011

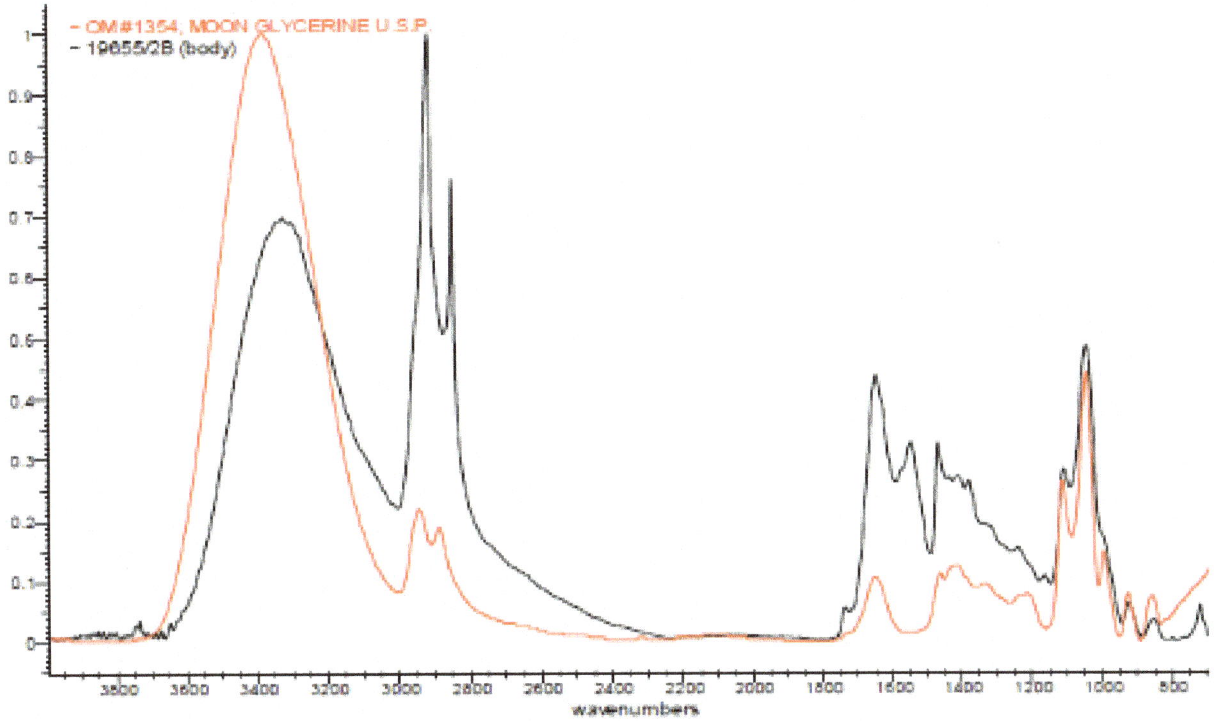

FIGURE 18 - Micro FTIR analysis of Ear Canal Specimen which matched Marvanol DP standard. Analysis performed by *Applied Consumer Services, Inc.*, Hialeah Gardens, FL ©2011

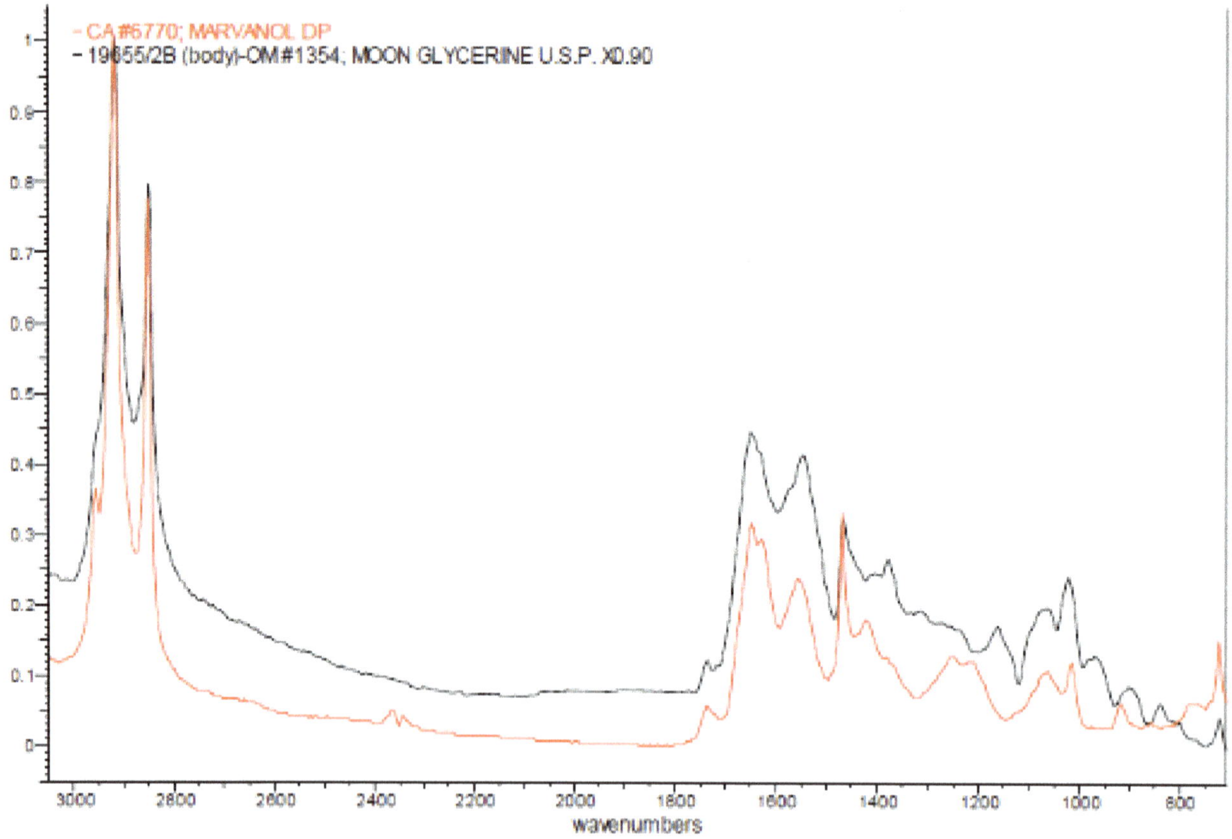

Name	Value	Unit
Name	MARVANOL DP	
Source of Sample	MARLOWE-VAN LOAN CORPORATION	
Technique	BETWEEN SALTS	
Solution Data	pH= (ON DILUTION) 3.5-4.5	
Comments	EASILY DISPERSIBLE IN WARM WATERChemical Description= SURFACTANTDescription= SOFTENER; CATIONIC ACTIVITY 15%	

RESULTS AND OBSERVATIONS

 Sample 19655/2A black square particles did not produce a response using FTIR. It was analyzed using dispersive Raman spectroscopy with a 532 nm laser and was determined to be highly consistent with graphite.

 Sample 19655/2B body mass was determined to be a mixture of two compounds. One was highly consistent with glycerine; the other, the surfactant *Marvanol DP*. This portion of the spectrum is consistent with a long chain amide polymer.

TABLE 1

Thorough Melting Point test on the "Brain Chip" of a Reactive Nematic Mesogen specimen from a Brain Tumor (Menginoma) specimen collected by *Integrative Health SystemsTM, LLC* on November 7, 2011, from the *Pathology Department* at *Cedar Sinai Hospital*, Los Angeles, CA as securely transported by *DiMatteo and Associates*, Moreno Valley, CA. The Melting point test was performed by *Applied Consumer Services, Inc.*, Hialeah Gardens, FL.

RESULTS

Temperature (Degrees Celsius)	*Observation*
300	No Change
375	Darkened and partially carbonized
400	Almost all the sample is black
450	Starts burning out
500	Powderizing
750	Hard powder
1000	Shrinkage, turning green
1100	Turns into hard green powder
>1100	No Change

PHOTOMICROGRAPH 1B - Spherical mesogen showing layers came from eye tear duct. Images taken by *Applied Consumer Services, Inc.* Hialeah Gardens, FL ©2011.

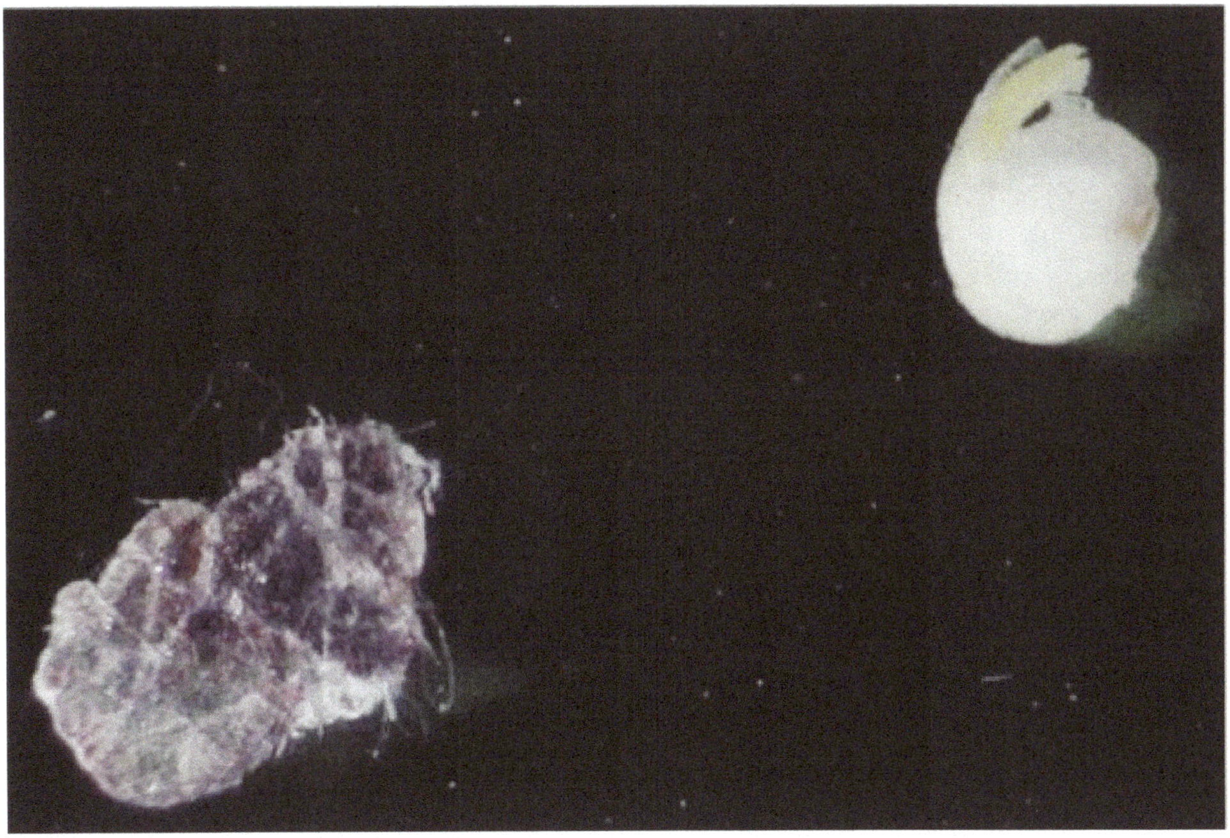

PHOTOMICROGRAPH 2 - Images of various types of Lyotropic Liquid Crystals and Metatrophic Liquid Crystals taken from left arm skin surface. Images taken by *Applied Consumer Services, Inc.*, Hialeah Gardens, FL ©2011

L/N 19825/2 ARM LEFT MESOGEN

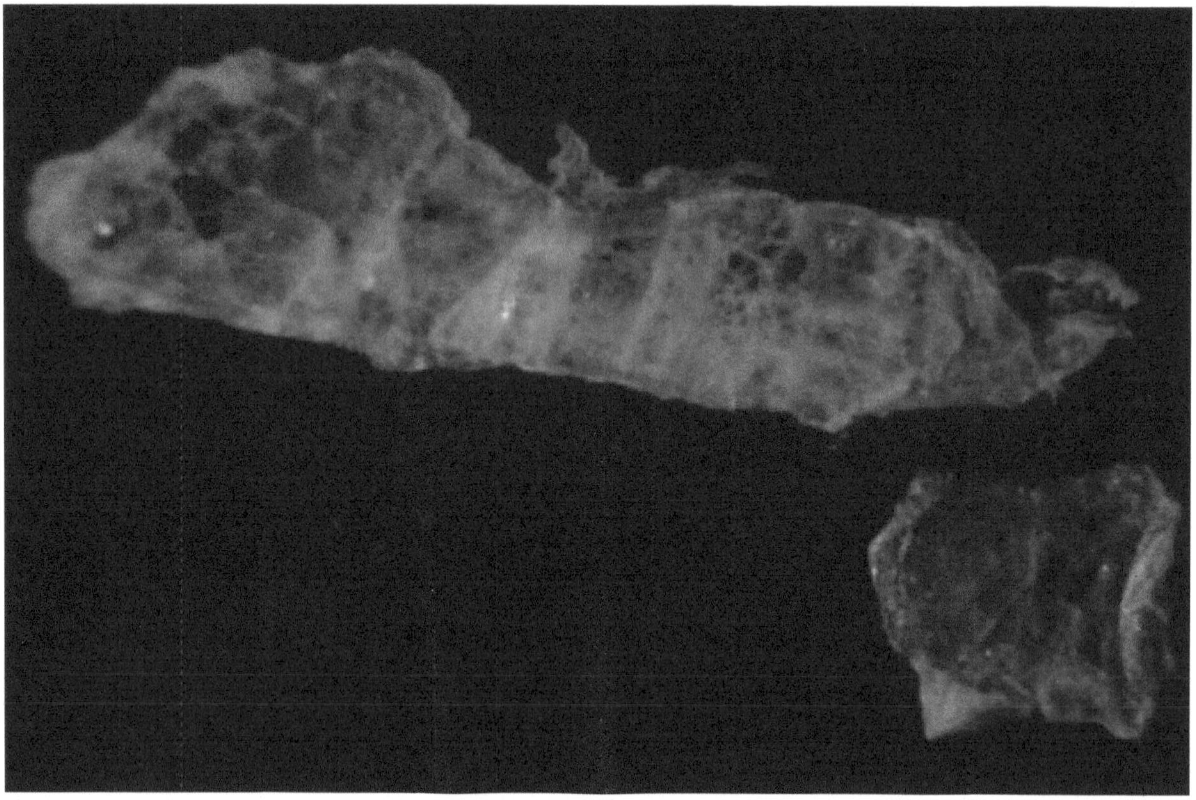

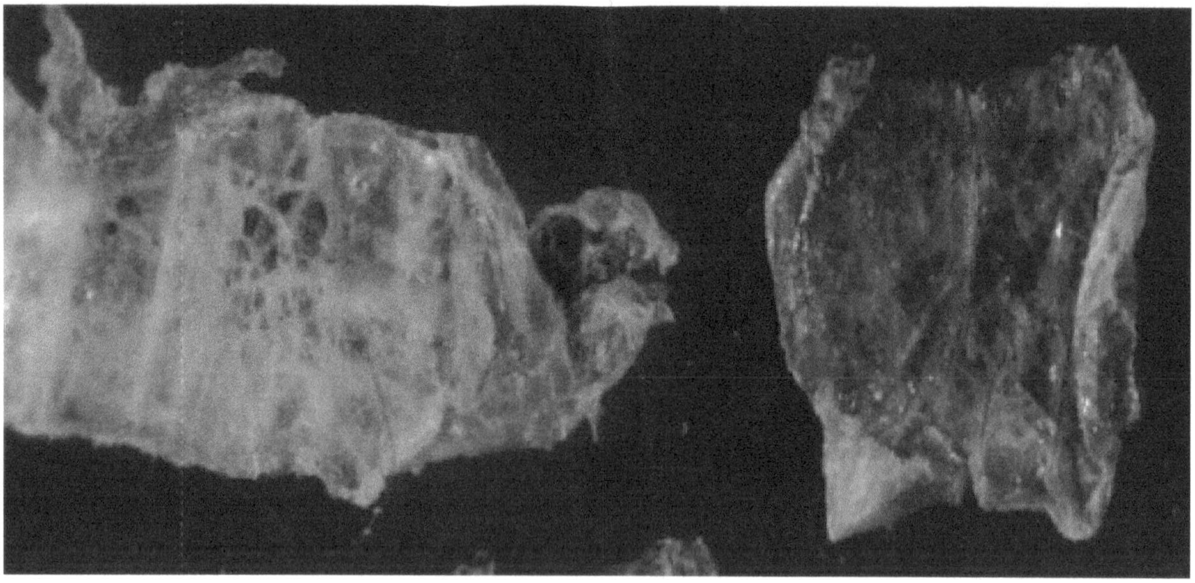

PHOTOMICROGRAPH 3A - Mesogen with directed energy thin coating. Note the standard liquid crystal mesogen in center. Image taken by *Applied Consumer Services, Inc.*, Hialeah Gardens, FL ©2011

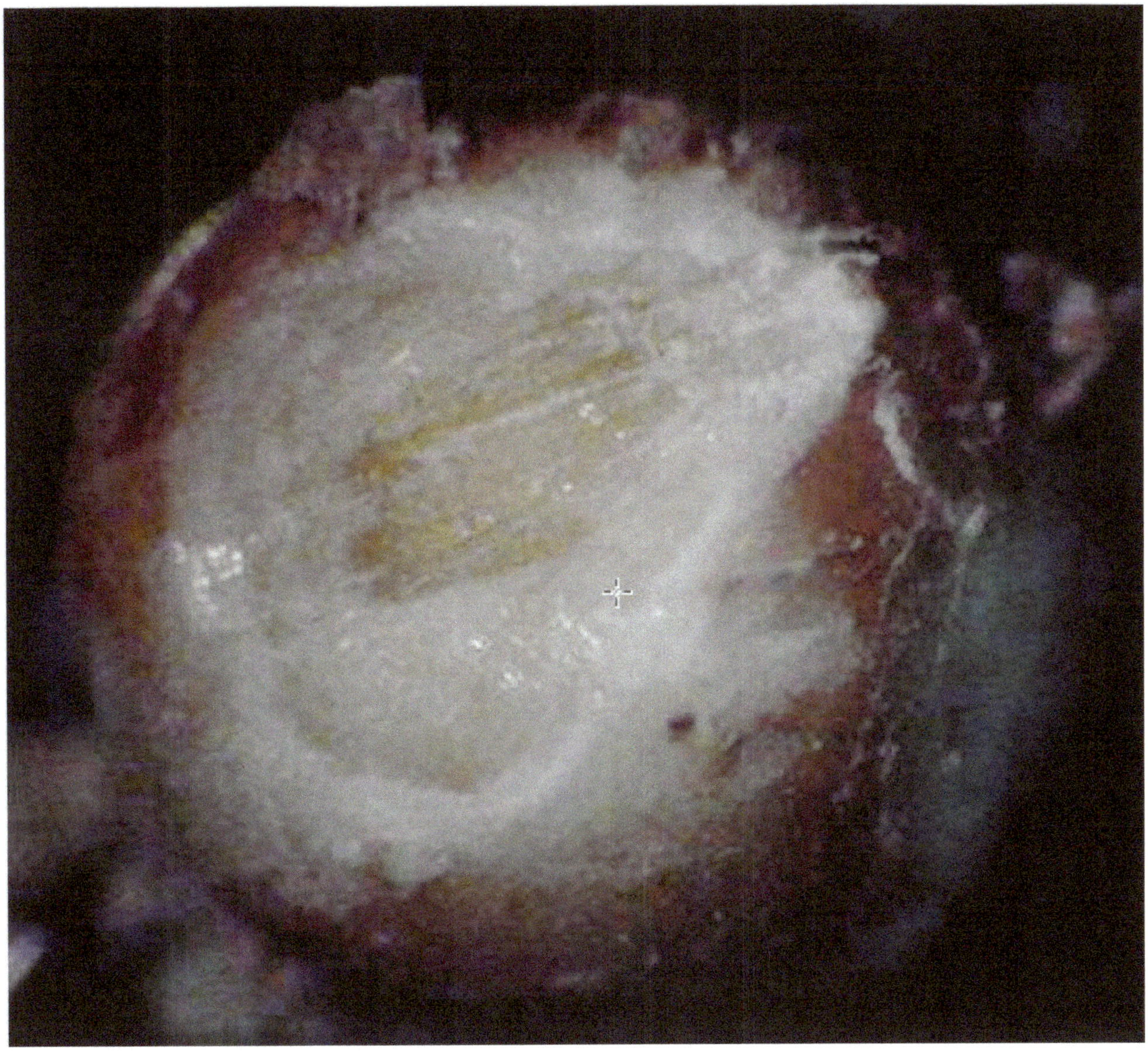

NASAL BULB SPECIMEN – NEW YORK

NASAL BULB SPECIMEN – SAUDI ARABIA

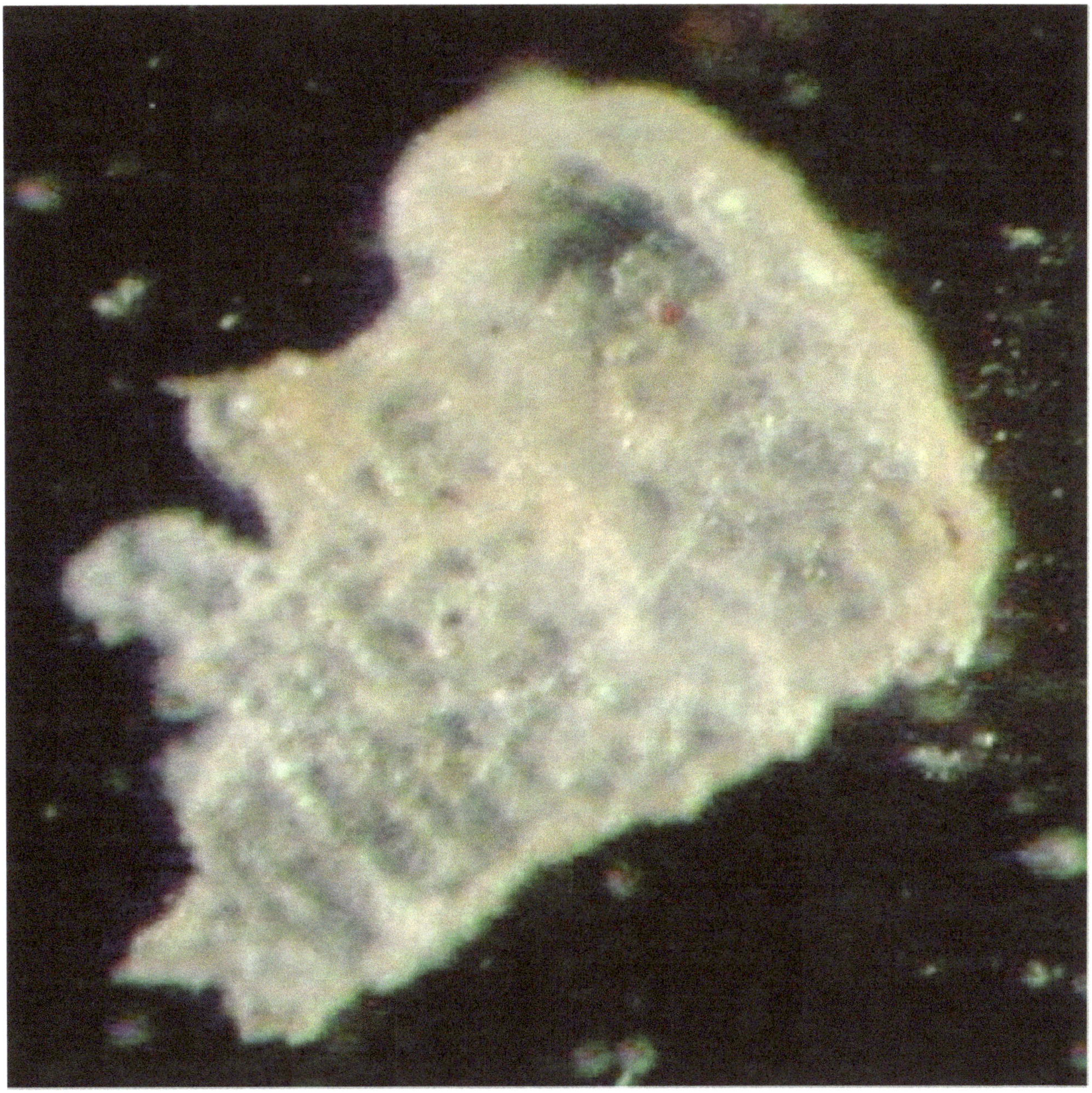

NASAL BULB SPECIMEN – LOS ANGELES

17733

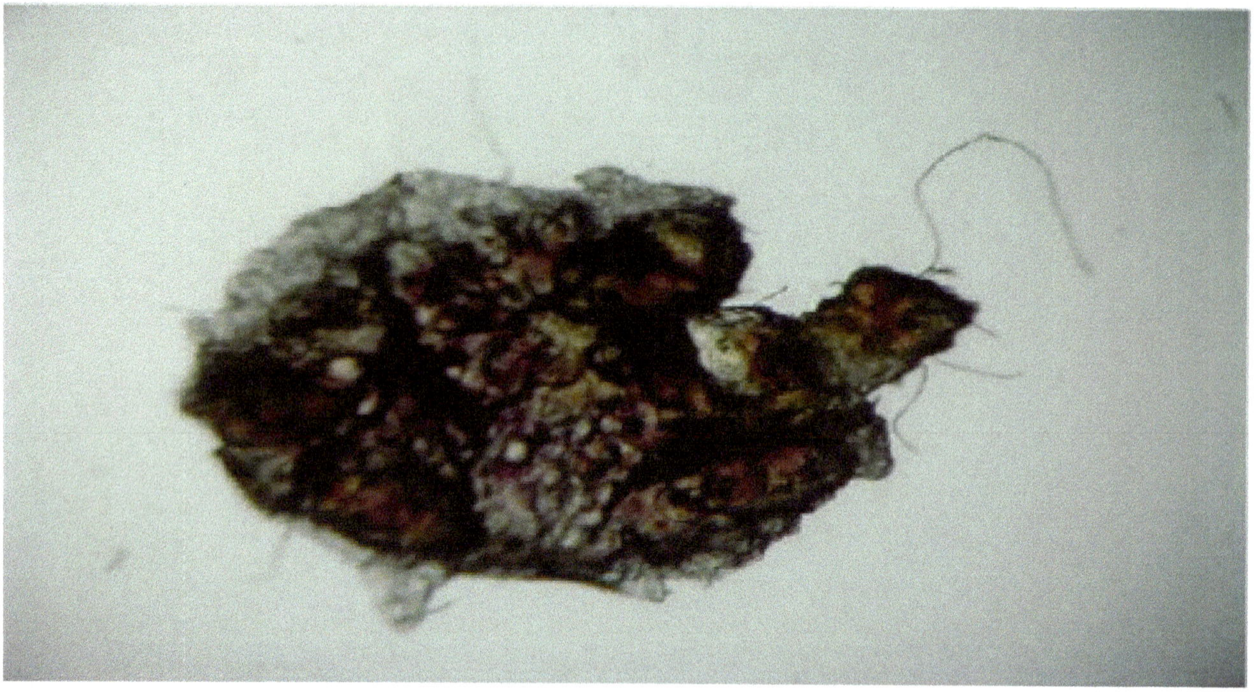

PHOTOMICROGRAPH 4 - Brain Menginoma ("Brain Chip"), reactive nemati mesogen. Image taken by *Applied Consumer Services, Inc.*, Hialeah Gardens, Florida ©2011

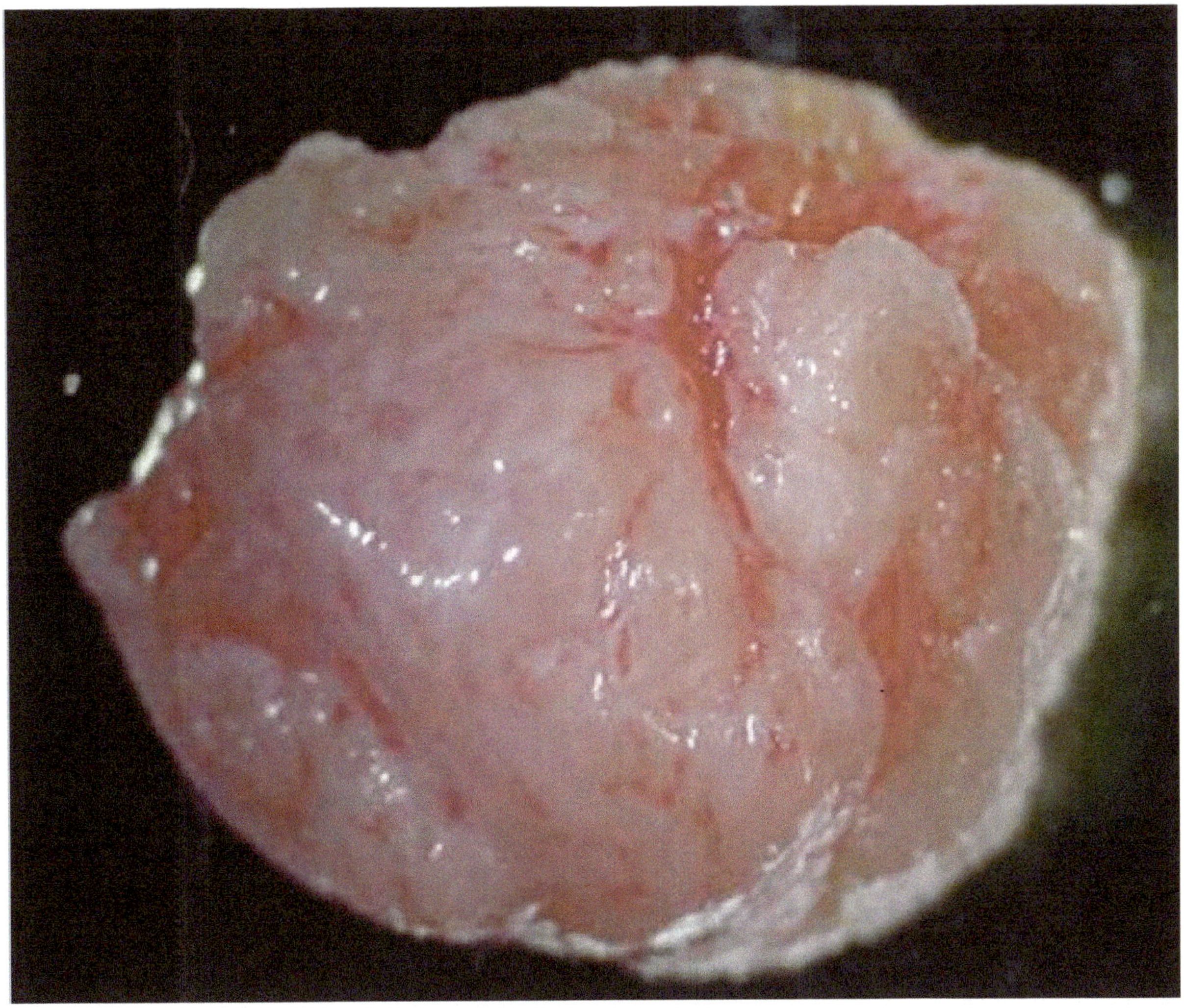

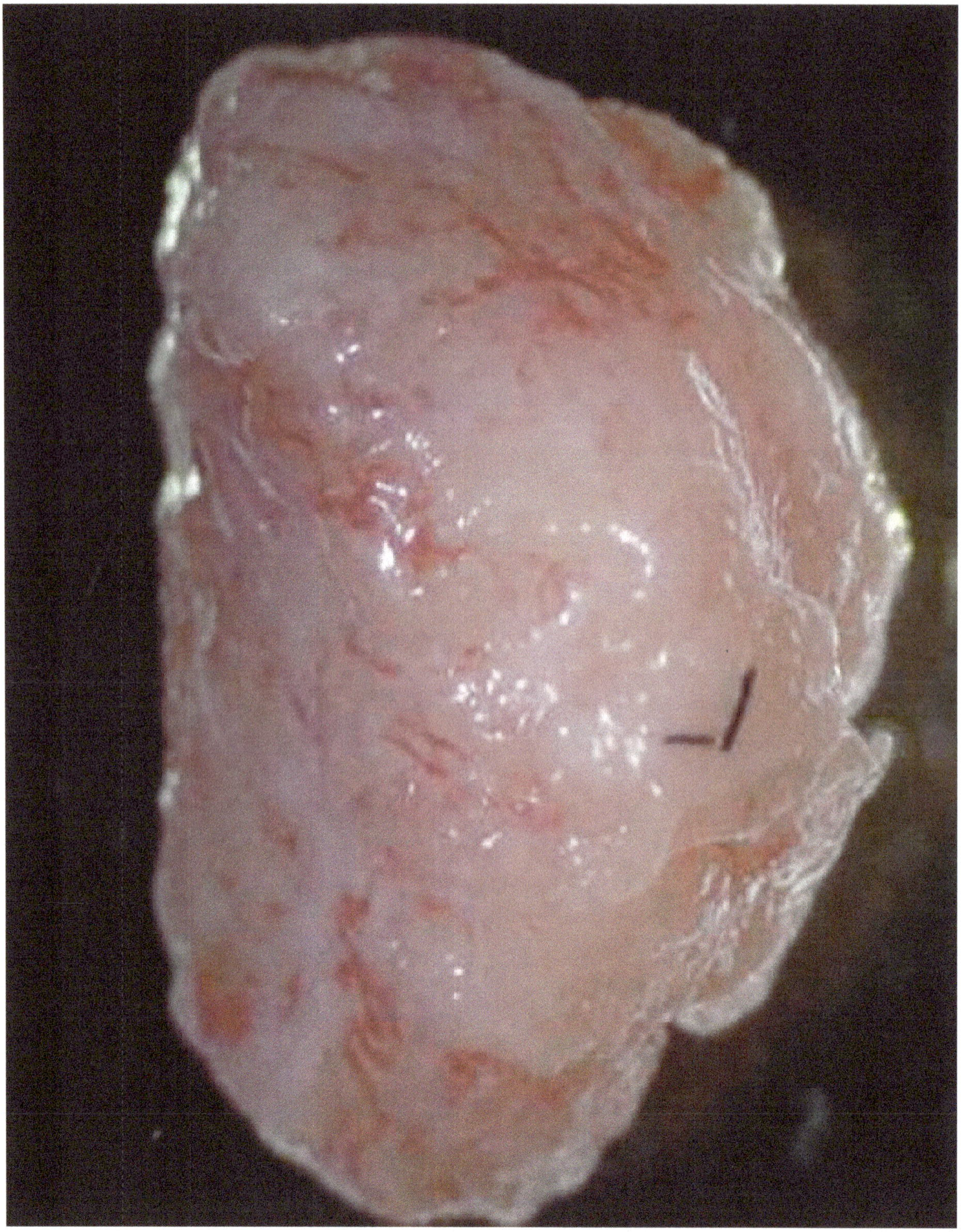

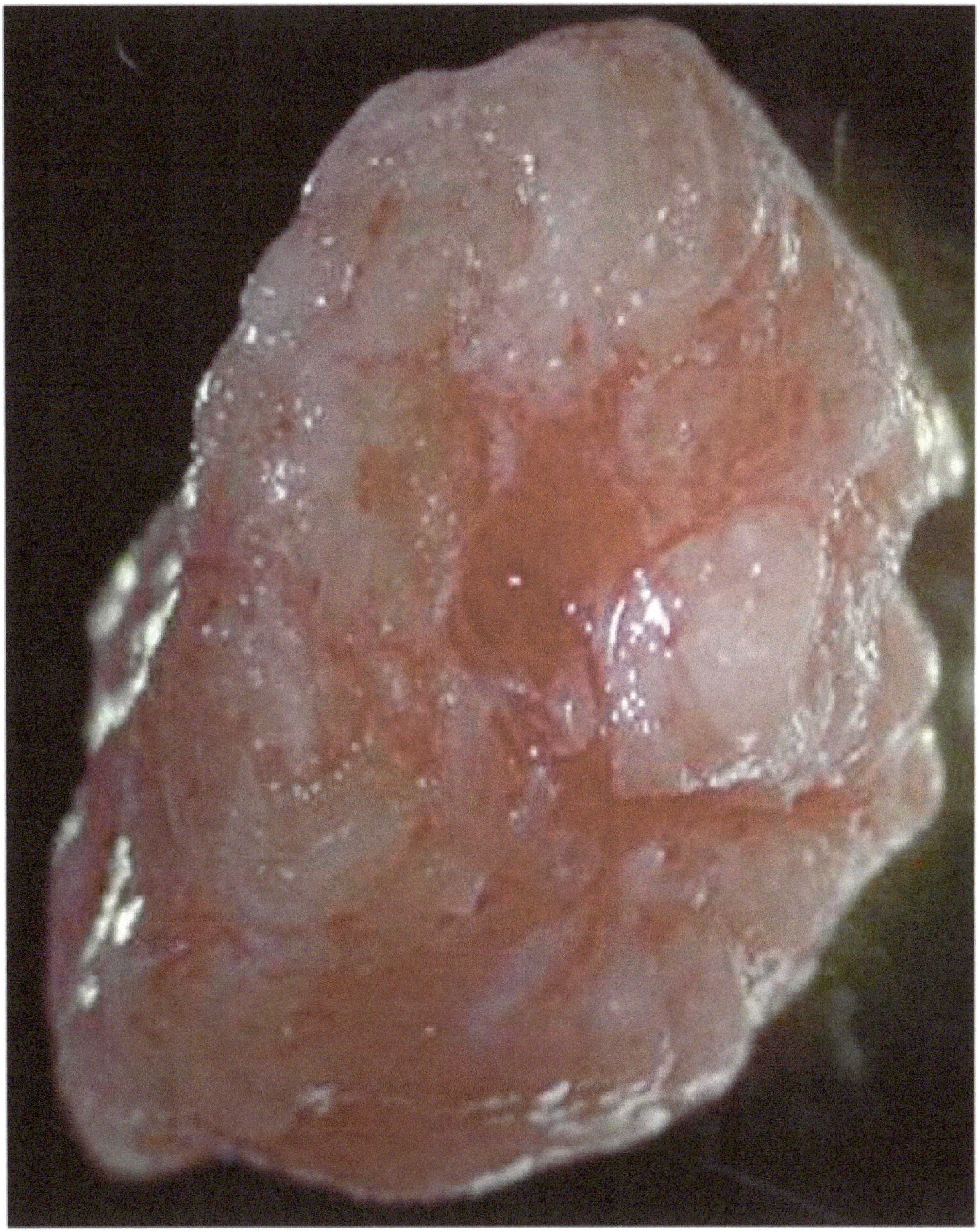

CUT SPHERE

(So hard that a diamond cutting saw had to be used)

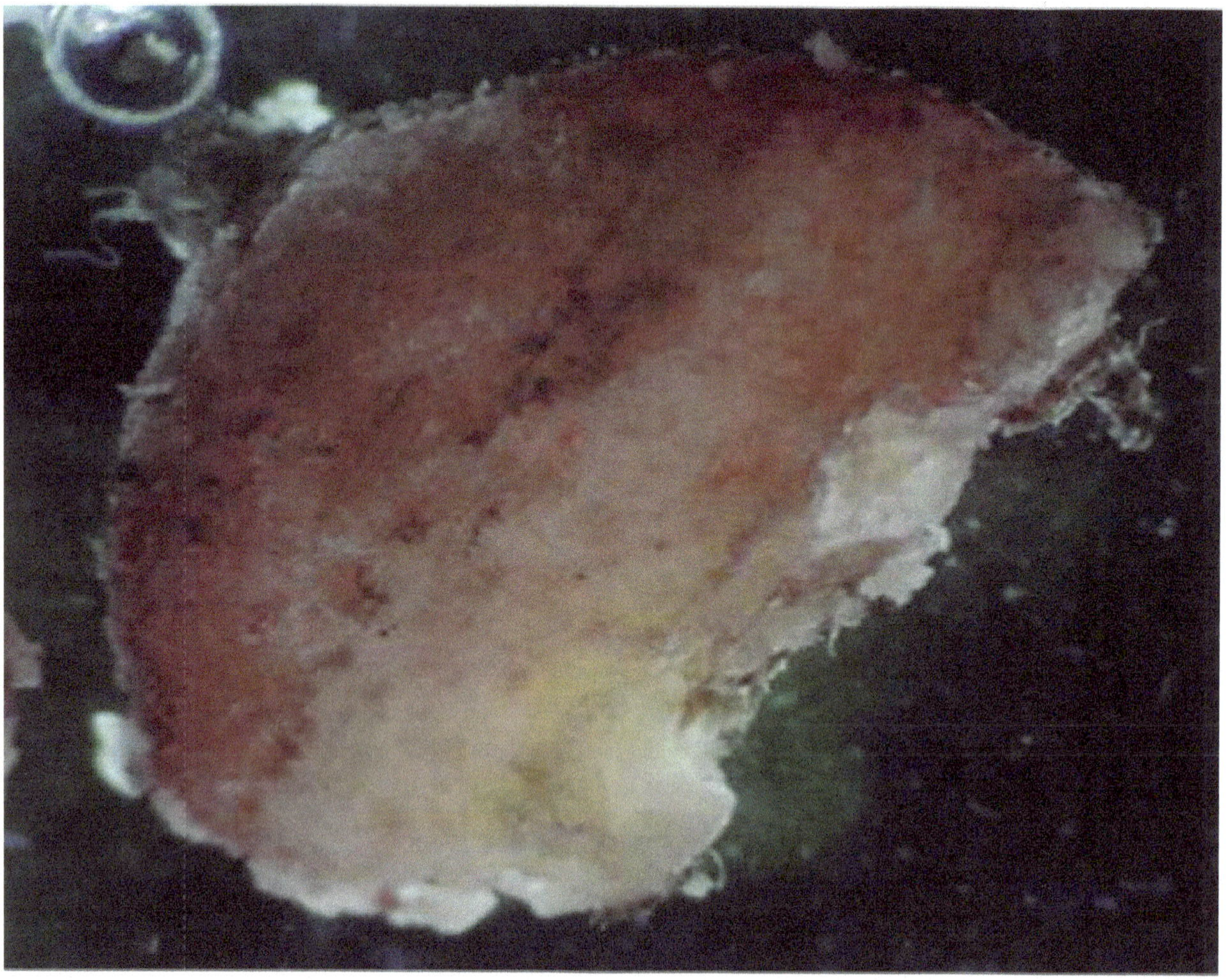

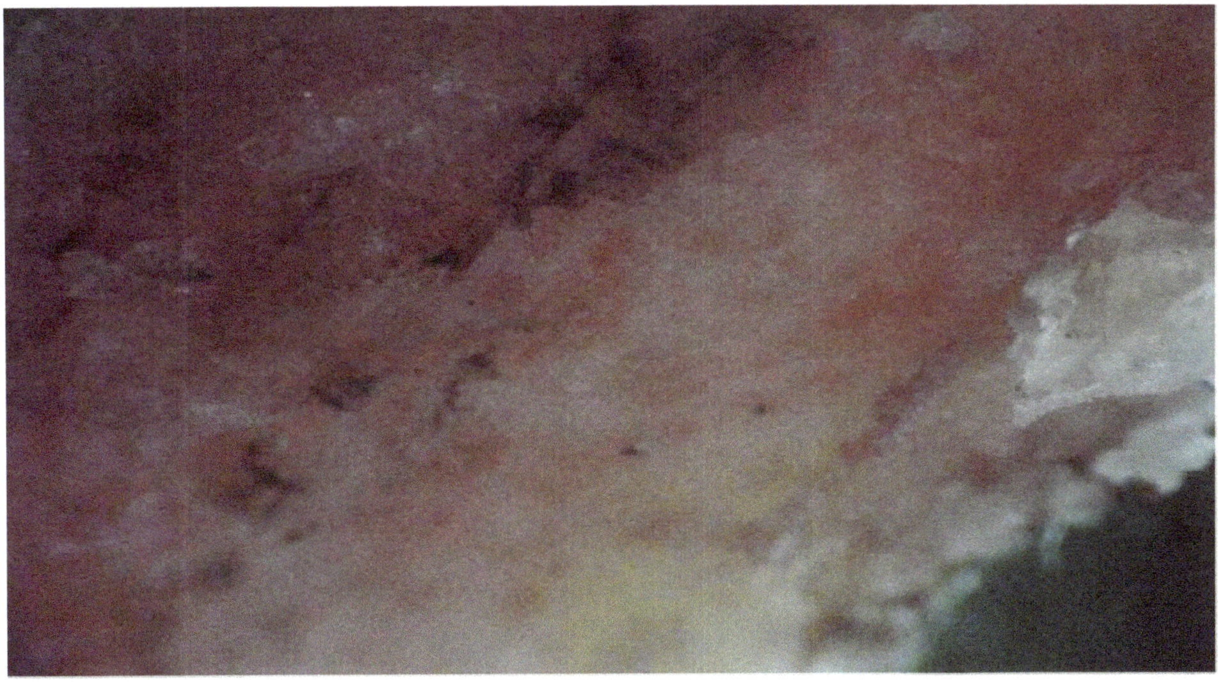

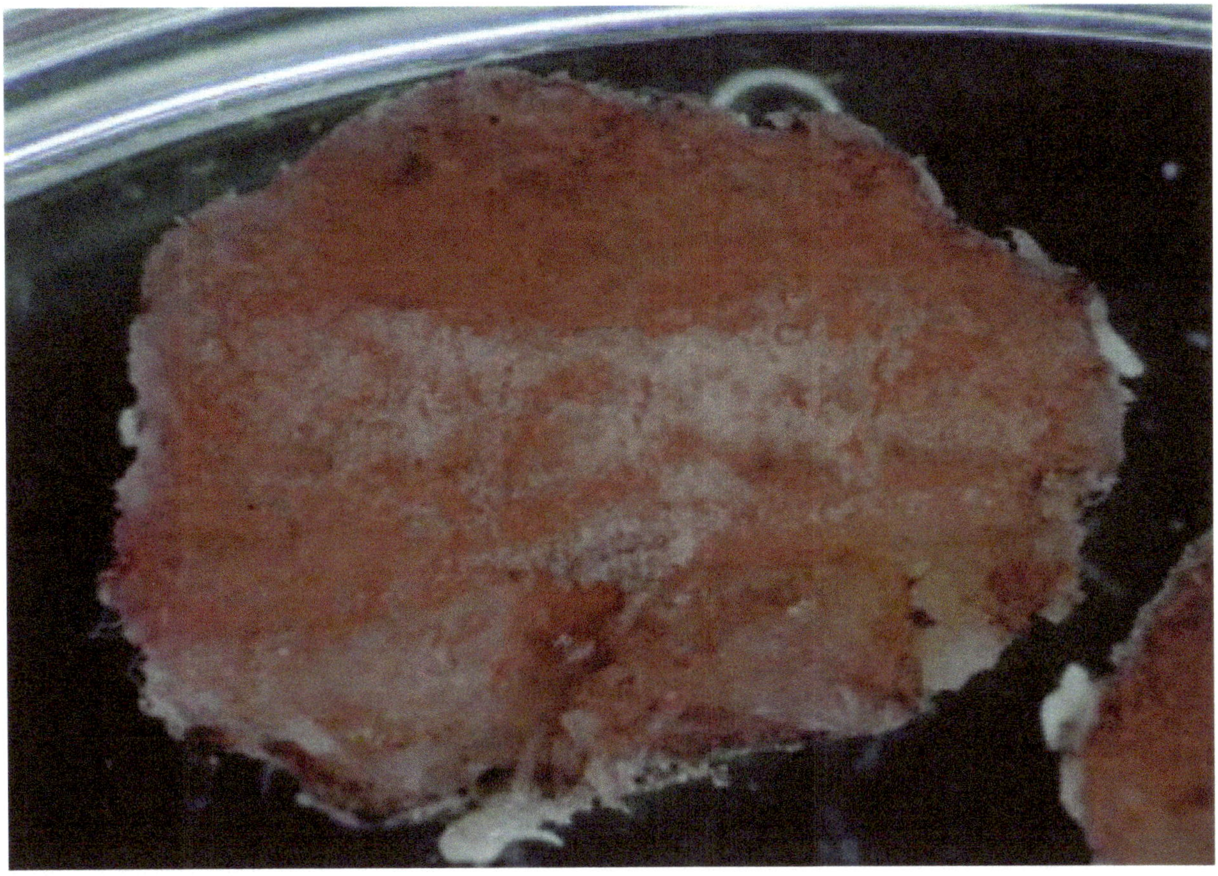

PHOTOMICROGRAPH 5 - Images of Ear Canal Specimens. Image taken by *Applied Consumer Services, Inc.* Hialeah Gardens, FL ©2011

EAR CANAL SPECIMEN A

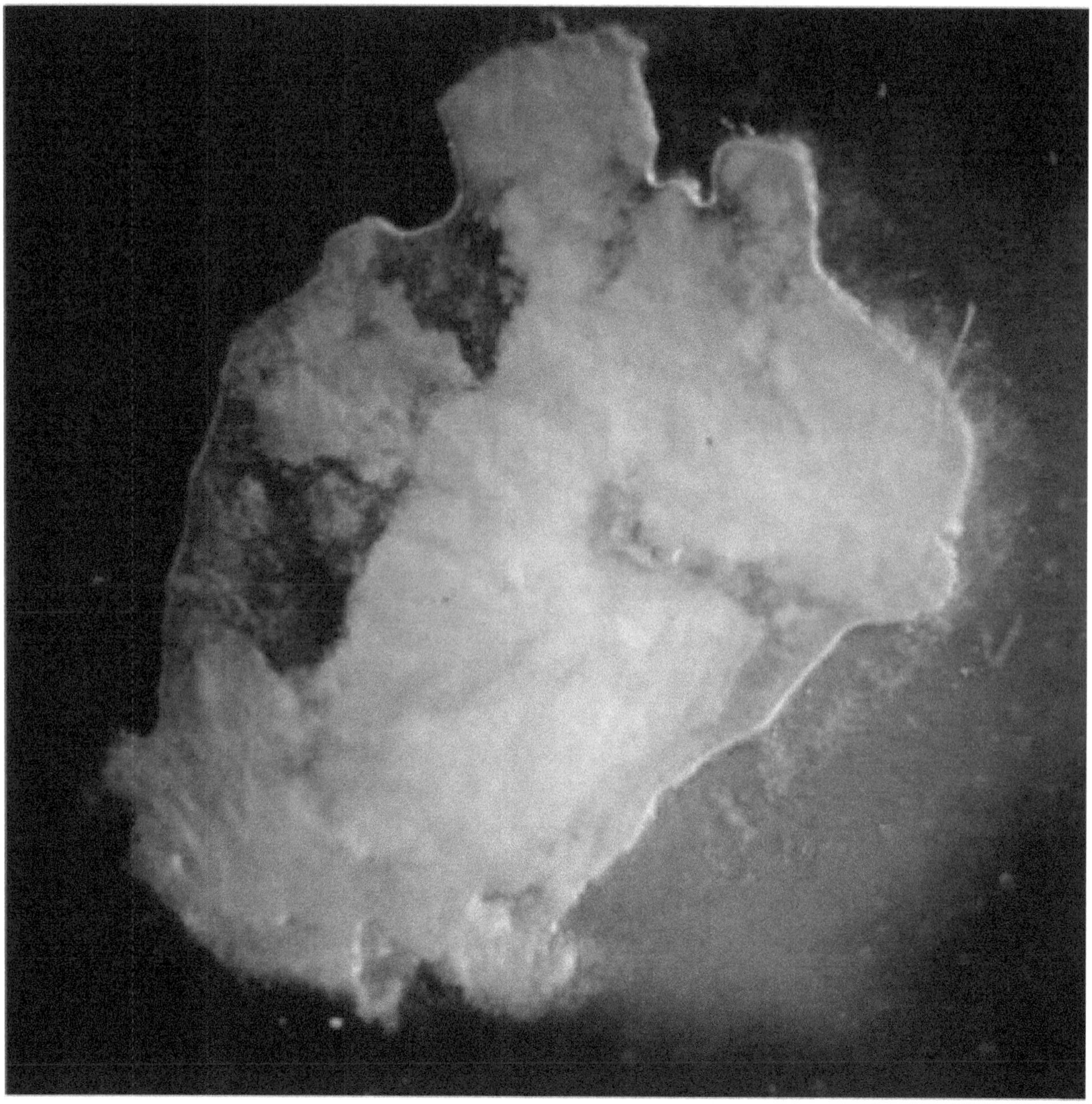

EAR CANAL SPECIMEN B

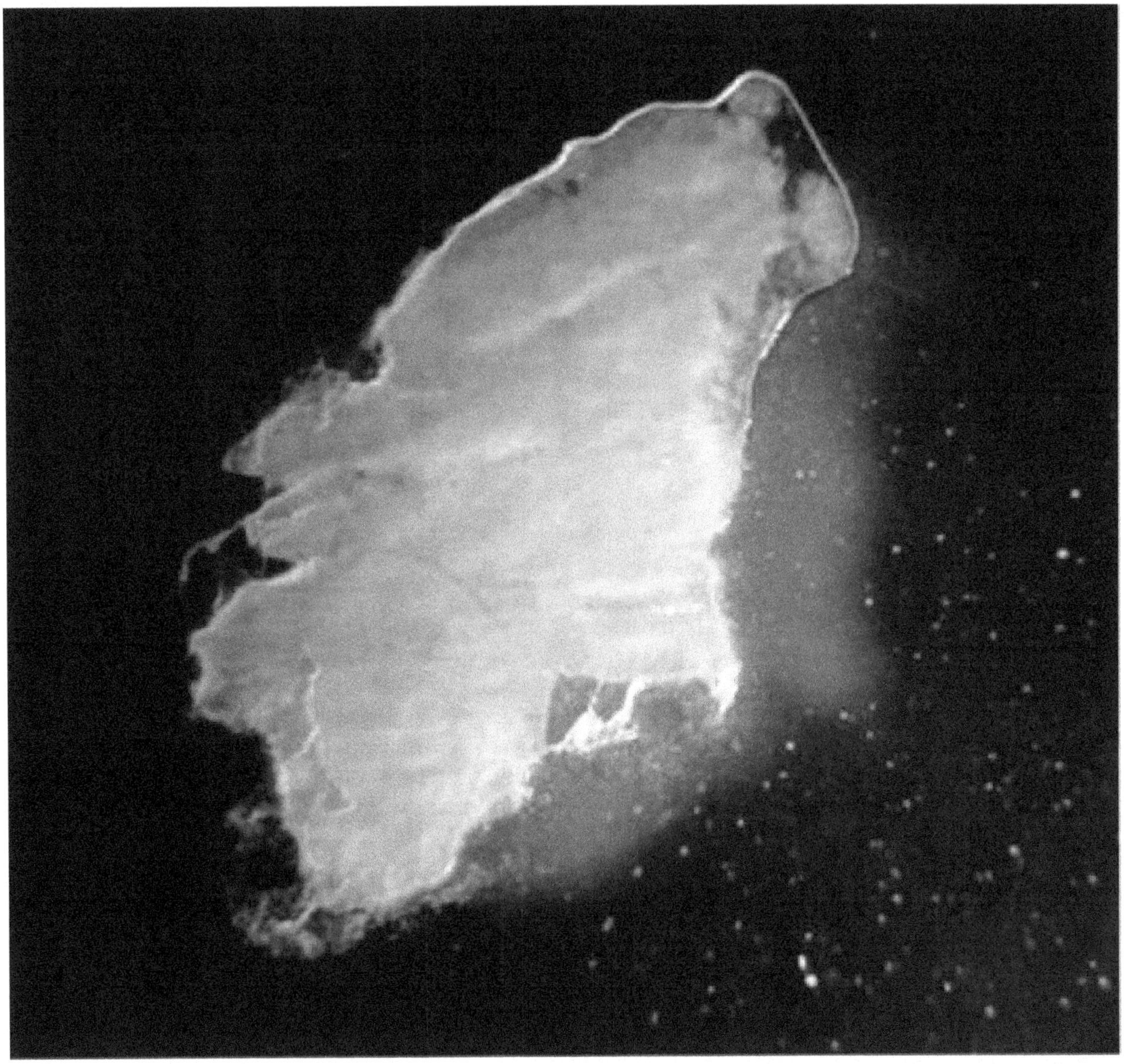

Ear Canal Wax Specimens

Note the fine nano materials and other materials. The black squares were determined to be graphite, which is a man-made compound. The material within the mesogenic nano building materials may actually be wax.

EAR CANAL WAX SPECIMEN A

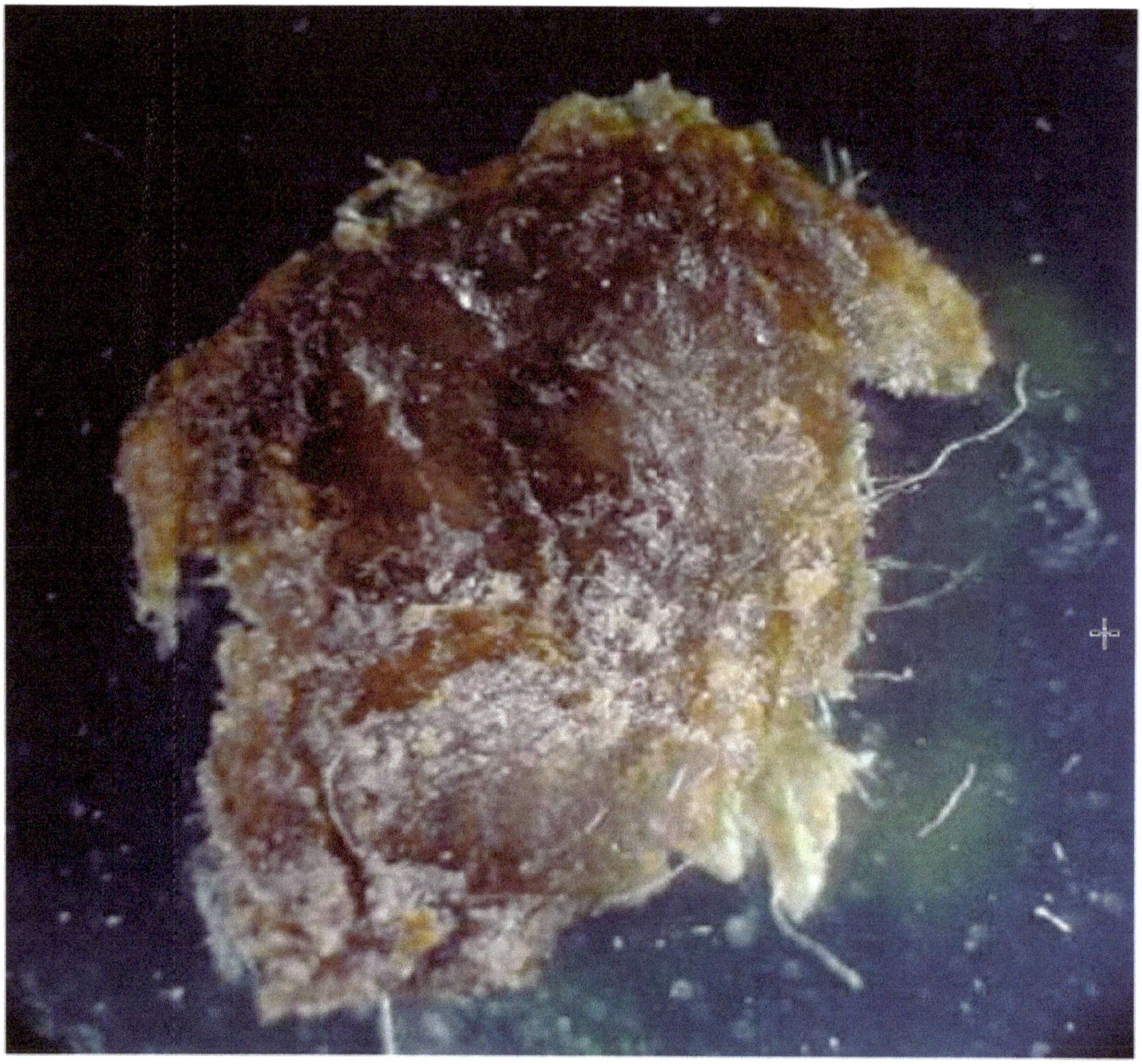

EAR CANAL WAX SPECIMEN B

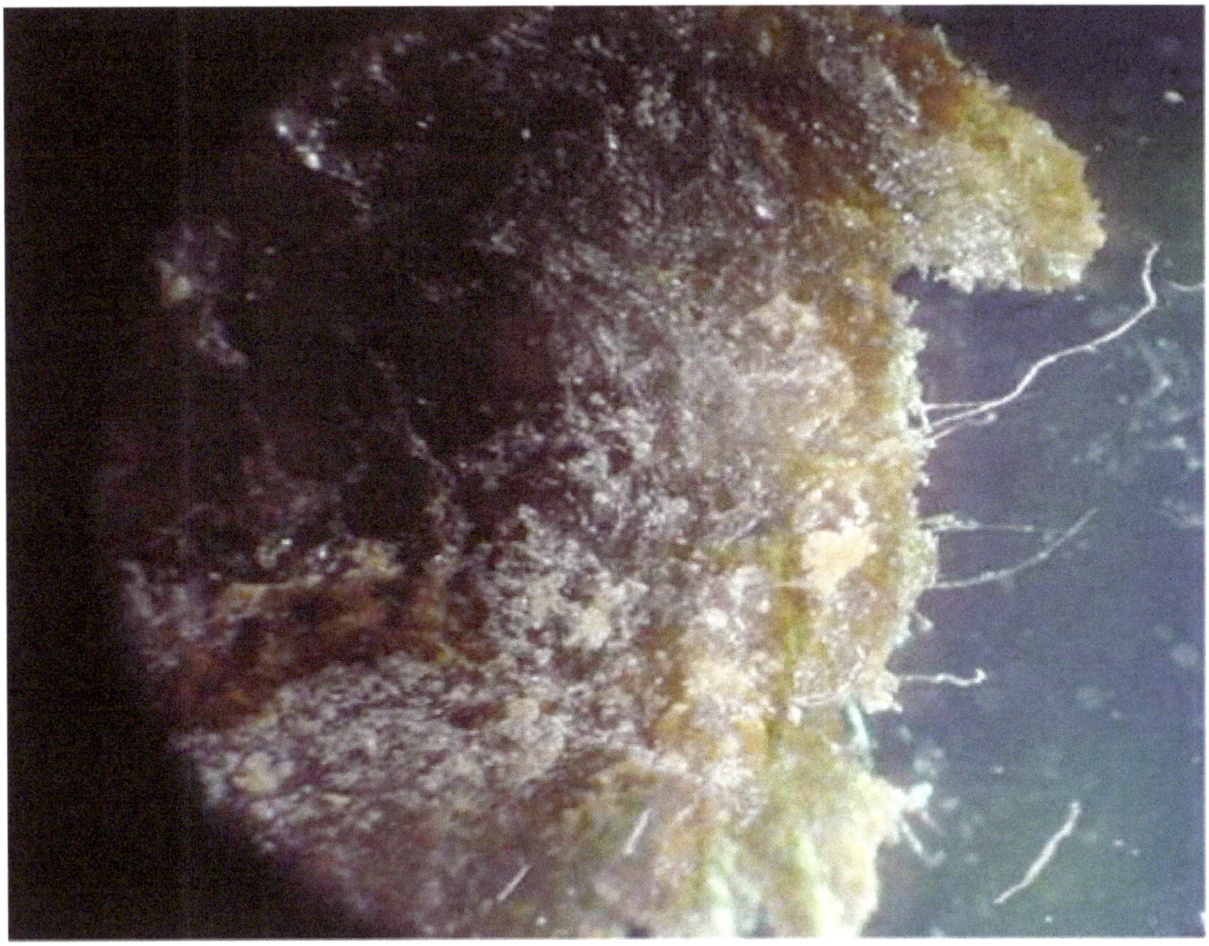

APPENDIX

The material contained in this volume text was submitted as previously published materials as referenced below with the appropriate "http.www" link by the original copyrighted and/or referenced site/author/publisher, except in the instances in which credit has been given to the source from which some to the illustrative and/or text material was delivered. These materials are to be utilized primarily for educational purposes and/or assisting in the professional development/training of professionals in various scientific, engineering, chemistry, allied health, medicine and other related disciplines, so they may become knowledgeable of the rapid State of the Art Advancement of Innovative Technologies.

Great care has been taken to maintain the accuracy of the information contained in this volume. However, Integrative Health Systems®, LLC, Science/Engineering Publishing Department: One Cell One Light® and/or Dr. Hildegarde Staninger®, RIET-1 cannot be held responsible for errors or for any consequences arising from the use of the information contained herein this Appendix Section and/or the volume text.

Materials appearing in this book prepared by individuals as part of their official duties as U.S. Government employees are not covered by the above mentioned copyright.

- Staninger, Hildegarde. SMART DUST: Defending the Land, Air and Sea and Its Global Environmental Impact through Depopulation Wars. Presented at the 2010 Health Freedom USA War Council™ Presented by Natural Solutions Foundation/Health Freedom USA. Webinar January 2 & 3, 2010 http://www/healthfreedomusa.org/?p=4327

- Wang, Zhong L. Ruiping P, Gao, James L., Gole, and John D. Stout. Adv. Mater. 2000, 12 No. 24, December 15 © Wiley-VCH Verlag GmbH.D-69469 Weinheim, 2000 0935-9648/00/2412-1938 $17.50 +.50/0

- Strukov, Dmitri B and Konstantin K. Likharev. A Reconfigurable Architecture for Hybrid CMOS/Nanodevice Circuits. Stony Brook University, Stony Brook, NY © 2006 (www.citeserx.its.psu.edu/viewdoc/download?doi=.1.1.7455&rep==rep1...

- Seo, Dongjin, Jose M. Carmena, Jan M. Rabey, Elad Alon and Michel M. Maharbiz. "Neural Dust: An Ultrasonic, Low Power Solution for Chronic Brain-Machine Interfaces. Department of Electrical Engineering and Computer Sciences and Helen Willis Neuroscience Institute. University of California. Berkley, CA © 2013 (www.arXIV:13072196v1 {q-bio.NC} 8 July 2013)

- Wu, Stephen S. and Marc Goodman. Science and Technology LAW: Neural Implants and Their Legal Implications. GP Sol-2013-January/February 2013: Practicing a Down economy-SCIENCE AND TECHNOLOGY LAW: Neural Implants and Their Legal Implications.

- (http://www.american bar.org/publications/gp_solo/2013/January_february/science_technology.law _neural_implants_legal_implications.html)

- http://phys.org/news/2016-07-americans-gene-brain-chip-implants.html

  "Americans worried about using gene editing, brain chip implants and synthetic blood" (PEW Research Center, July 26, 2016)

SMART DUST: DEFENDING THE LAND, AIR, and SEA and ITS GLOBAL ENVIRONMENTAL IMPACT THROUGH DEPOPULATION WARS.

by Hildegarde Staninger, Ph.D., RIET-1. Integrative Health Systems, LLC., 415 ¾ N. Larchmont Blvd., Los Angeles, CA 90004. Phone: 323-466-2599 Fax: 323-466-2774 e-Mail: ihs-drhildy@sbcglobal.net.

Presented at the 2010 Health Freedom USA War Council!™ Presented by Natural Solutions Foundation/Health Freedom USA.

Webinar January 2 & 3, 2010 http://www.healthfreedomusa.org/?p=4327

Abstract: The world is in the cataclysmic reactor of the universe where mankind is desperately trying to protect his nations through the advancement of science, medicine, and technology. And at the same time, the use of this bionanotechnology architecturally designed commercial products are at a fork in the road with the possibility of killing many of its inhabitants without the destruction of a single building. This is the time for all people to listen to their hearts and realize, that through the advancement of nanotechnology and its bridge to artificial biology, their greatest creation, "Smart Dust", will be the way something nearly invisible could become our greatest national security threat. Its use as a monitoring tool has taken many tolls on the global population of plants, animals, and mankind. Whether it was designed as a universal surveillance tool or the monitoring tool of the rate of death of a tree, it is here and is part of the global dominance race involving smart nanotechnology. This paper will present an historical overview and a pictorial overview of field observations of the technology as found in the human population of the United States of America.

Introduction to Nanomaterials and Nanotechnoogy

Nanomaterials are great and strange at the same time. Carbon nanotubes, for example, add strength, flexibility, and heat protection to plastics, ceramics, and metals. Nanomaterials don't break easily when dropped or smashed. When sliced, nanomaterials "heal" themselves by linking back together. Nanomaterials offer engineers a brand new bag of tricks to make life better for everyone.[1]

Nanomaterials have amazing and useful properties with many structural and nonstructural applications. But, they are not necessarily new –

nanomaterials have been important in the materials field for a long time; we just couldn't see or manipulate them. Gold nanoparticles were used in medieval stained glass, and nanoparticles of carbon black (graphene) have been used to reinforce tires for nearly 100 years.[1]

The advancements of nanotechnology in the last 10 years are like the early chemistry, or *alchemy,* as it was known. Alchemy's greatest day was during the time of the power plays of monarchs, like Philip II of Spain and his nephew, who were the greatest European contributors, supporting the science of alchemy, astrology and mathematics through the use of Philip II's "*magic circle*". Rudolf II of Hapsburg, during the time of Renaissance Prague[2], within the walls of his Prague castle, contributed to the world by supporting the works of Tycho Brahe, Johannes Kepler, and the artist Albrecht Durer, to name a few. He allowed and financially contributed to the birth of chemistry as a modern-day science. Through these early foundations of chemistry, the promise of amazing things to those who understood its power would be given, just as nanotechnology will give the greatest commercial economic rewards to the companies and nations controlling its architecture and deployment.

Nanotechnology is expected to have an impact on nearly every industry. The U.S. National Science Foundation has predicted that the global market for nanotechnologies will reach $1 trillion, or more, within 20 years. Nanotechnology-related job projections are estimated to be at nearly TWO million engineers/scientists, worldwide, by 2015. Working in Nanontechology offers significantly more rewarding career and job satisfaction compared to mundane careers in IT or similar fields. However, there is little awareness or knowledge in India about world opportunities in Nanotechnlogy. Nanotech training is determined to bridge this gap and avail world-class knowledge and opportunities in Nanotechnology to students and professionals in the world. There are currently over 997 global universities involved in research and the commercialization of nanotechnology with industry, governments, and military. Many of the world's leading Nanotechnology Institutes and companies, have developed training courses in Nanotechnology. The aim of nanotech training is to open a window to immense activities and opportunities in Nanotechnology, to guide and eventually realize a dream of launching a coveted career in the next generation technology known as NANOTECHNOLOGY -- but at what costs to the environmental and humanity?[3, 4]

To begin our journey into the world of NANOTECHNOLOGY, fellow scientists, engineers, and medical professionals who have not kept up with the state-of-the-art developments in this rapidly growing field, need to understand the fundamental building blocks used in the architecture of nanomachines, tools and sensors used as our new surveillance tools for global safety and monitoring of natural and manmade pandemics.

Carbon Nanotubes

Single-walled carbon nanotubes (SWNTs) are incredibly promising nanomaterials. Their remarkable materials properties, such as strength, rigidity, durability, chemical vigor, thermal conductivity, and (perhaps most importantly), electrical conductivity, make them very versatile. Depending on their exact molecular structure, some nanotubes are semiconducting, while others display true metallic conductivity. This ability, combining with their nanoscale geometry, makes them great candidates for wires, interconnects, and molecular electronic devices.[5]

Through the development of super acids, as discovered by Professors Richard Smalley and Matteo Pasquali at Rice University's Center of Biological and Environmental Nanotechnology (CBEN), such as sulfuric acid, work well in dispersing SWNTs into an easily processed form. This carbon form ranges from individually dissolved nanotubes, to a liquid crystal which acts as a starting material for aligned SWNT fibers. This improved material, then, has laid the ground work for larger objects (nanomachines and tools) made entirely of SWNTs.[5] Figure 1-1 shows self-assembled carbon nanotubes in a simple diagram. Figure 2-2 shows them as in an actual field specimen from a female exposed to advanced nanomaterials (Project: FMM © 2006 H. Staninger, Photomicrograph taken by Dr. Rahim Karjoo, Pathologist).

Multi-walled carbon nanotubes (MWNTs), with an average diameter of about 40 nm, also have a variety of potential uses in everything from cell phone lens systems and shutter materials to car windows and sporting goods. Their comprehensive strength appears even greater than SWNTs, as is proving important in composite materials. Recent developments by Massachusettes Institute of Technologies, Soldier Institute, in development of the "nanoworm" (which is able to take pictures of cells within the body and transmit them back to a computer), are the highlights for modern nano medicine in 2007.[6] Close review of Figure 3-3 shows a picture of an advanced nanomaterial device called "Goldenhead", which originally, upon analysis, had a layer of high density polyurethane (a simple lens) (wall No. 1). It had a second wall covered with acrylonitrile/acrylic resins with specific dye impregnated into its self- assembling polymer composite material, along with a viral DNA sensory camera at its end, seen as the red dot. (Taken from Project: FMM, © 2006 H. Staninger, Photomicrograph taken by Dr. Rahim Karjoo, Pathologist).

These tough technical puzzles and fabrication obstacles are only the beginning in the area of pathogen countermeasures, which looks at ways to counter exposures to populations. In many of these types of technologies (due to the metallic conductivity of the SWNTs and MWNTs materials), one may be able to combine these materials with coating materials such as paints, dyes, and adhesives. When mixed with paint, nanotubes become electrostatic. This helps plain and other nanotube-containing coatings stick more tightly to surfaces. Cars on the production line could be coated with nanotube pigments to cut manufacturing costs. Nanotube pastes might be able to improve liquid crystal and flexible displays to create images sharper than silicon and carbon-based films can generate. When these technologies are coupled with specific frequencies, such as 288 MHz, one can send a wireless image to the advanced nanocomposite materials. This can be an image of some horrible, created disease that does not exist - just an image upon nanoclaws, hooks, wires and sol-gel. (See Figure 4-4, image of self-developing nanocomposite material after 21 days from removal of human source. Project: FMM © H. Staninger, photomicrograph taken by Dr. Rahim Karjoo).

Nanocrystalline Materials

All of us who ever took a single science class know that everything is made up of atoms and molecules. Most of these bulk materials have particles varying in size form hundreds of microns to a few millimeters.

Nanocrystalline materials have particles of about 1 to 100 nm. An average atom is about 1 to 2 angstroms (A) in radius. One nanometer is roughly 10 angstroms (A), and there may be three to five atoms in one nanometer, depending on the size of the atoms.

Nanomaterials are particularly strong, hard,ductile (bendy/stretchy) at high temperatures, wear, erosion and corrosion resistant, and chemically reactive. Many departments may be addressed, depending on their various specific properties. For example, nanosilver has special catalytic properties that bulk silver does not have (e.g. interacting with and killing virus).[7, 8]

The following lists of methods are commonly used to produce nanomaterials:

- Sol-gel (colloidal) synthesis.
- Inert gas condensation.
- Mechanical alloying or high-energy ball milling.
- Plasma synthesis.
- Electrodepostion.
- DNA folding Origami Proteins.

Although all of these processes are used to create various amounts of nanomaterials, currently sol-gel synthesis is able to:

- Make precision materials in large quantities, fairly cheaply.
- Create two or more materials at the same time.
- Make extremely homogeneous (same throughout) alloys/composites and ultra-high purity (99.99 percent) materials.
- Produce materials (ceramics and metals) at ultra-low temperatures (around 150 to 600°F compared to 2500 to 6500°F in standard methods).
- Fine-tune atomic composition/structure accurately.
- Couple with synthetic DNA/RNA to create new vaccines or genetic repair nanomedicine delivery systems.

By creating or augmenting materials at the nanoscale, applications engineers can add capabilities such as superior strength to existing products.[8, 9]

Nanolasers and Nanocrystals

All forms of modern communication systems, whether on an airplane or a satellite system of advanced nano materials, a *repeater*[10] is necessary for in-line amplifiers that take fading phonophotonic signals and resend them with more power. To do that with data shot along the line as well-ordered groups of photons and phonons, those repeaters need to be miniature coherent-light sources called nano-size lasers.[11]

The photonic band gaps created inside photonic crystals not only provide an excellent way to keep the photons moving along certain paths, they can also provide areas that trap photons – *optical cavities.* As light enters such a cavity, the photonic band gap (etched into the crystal) keeps the light from leaving through the rest of the crystal. Such "trapped light" bounces back and forth in the cavity – gaining in energy, tightening into a coherent beam.

Certain materials – some semiconductors for example – can be stimulated to emit photons. The LED (light-emitting diode) is a common example of this kind of technology, but is not a laser in itself. To get the laser effect, you would most often want to choose a semiconductor-like material – called a *gain medium* – and place it in an optical cavity. Photons enter the cavity, bounce back and forth, and stimulate the medium to emit more photons. Those extra photons are of the same wavelength as the original ones, so what you get is an amplified version of the original light. (This "light amplification by stimulated emission of radiation" is how a laser gets its name.)

(See Figure 5-5: Nano light-emitting diode from Sencil™ technology © Integrative Health Systems, LLC photomicrograph Applied Consumer Services, Inc.)

To make a nanolaser, a photonic crystal is used to create a cavity that is almost as small as the wavelength of the photons themselves. This cramped space forces the photons to travel in nearly parallel lines, until the intensity of the light reaches the theoretical limit - in effect, all the photons are traveling right on top of each other! The gain medium is essentially part of the crystal itself – but before it can emit photons, a small electrical current must be induced (this is true for most semiconductor lasers), through the use of a gain medium – like chemical in the silane family or a nanopiezo electrical device.[12, 13, 14] (See Figure 6-6: Silane Based Nano Composite Material (2 polymeric materials and silicone head; Figure 7-7: Acrylic Nano Composite Material as a Piezo Electrical Device and Figure 8-8: Edible Braille RDIF Chip as compared to granules of sugar.)

The little zap of extra energy is all the photons need to make a break for it and blast out of the crystal as a laser beam. When supplied with a little electricity and a signal composed of photons (like a light house sending its signal to the captains of the sea), the laser amplifies that signal. Out comes the flow of happy photons, ready to spread information around the world or even to a cell in the brain.[15]

Korean researchers have developed such a photonic crystal laser, using semiconducting materials (indium, gallium, arsenidem and phosphide). Their laser produces detectable amounts of light with as little as 250 millionths of an ampere of electricity. Their design uses a tiny post (nanoanchor) to conduct electricity and soak up excess heat without disturbing the main portion of the crystal at the top.[16]

Self-Assembling Crystals

Nanocrystals are clumps of atoms that form a cluster. They are bigger than molecules (~ 10 nm in diameter), but not as big as bulk matter. Although nanocrystals' physical and chemical characteristics change, one of their big advantages over larger materials is that their size and surface can be precisely controlled and properties tuned, like quantum dots (a type of nanocrystal). In fact, scientist can tune how a nanocrystal conducts charge, decipher its crystalline structure, and even change its melting point.[17]

Dr. Paul Alivisantos, chemist at the University of California at Berkeley and Lawrence Berkeley National Laboratory has made nanocrystals by adding semiconductor powders to soap-like films, called surfactants. His group has grown mixtures of crystals using different surfactants. These react with semiconductor powders and produce different-shaped nanocrystals (e.g., rods instead of spheres).

Dr. Alivisantos's ability to grow semiconductor nanocrystals into the shape of 2D rods opens up many new applications and shows how controlling crystal growth as important to changing size and shape. Although shape change is at its horizon of being understood by fellow nano-ologists, it is possible that the interaction of atoms with different surfactants causes a crystal to grow in a particular way – just as various hazardous materials trapped in various tissue pockets will react to the administration of nanocrystals into the human body.[18, 19] To keep up with a speedy growth rate, then (with the right mix of surfactants), crystals take on elongated, rod-like, and faceted shapes, such as nanohorns as made with the mixture of waterborne polyurethane (styrene) and high-density polyethylene, which yields a styrene-based plexi-glass of nanocrystals growing out of the skin of an individual. (See Figure 9-9)

Further studies by Dr. Alivisato's group have shown that rod-shaped nanocrystals give off polarized light along their long axis compared to nonpolarized light fluoresced by cadmium selenide nanocrystal spheres. This is important in biological-tagging of synthetic biology, stem cells, and viral protein envelopes.[20] Since nanorods can be packed and aligned (like logs on a rail road car) they may also work well in LEDs and photovoltaic cells.

Dr. Alivisatos and others have shown how they can change nanocrystal growth conditions and rates to create nanocrystals in the shape of teardrops, arrowheads, "jacks," and horns. These shapes, since they are so small, are registered with the US Department of Commerce under the term "molecular brand,"[21] just as the continuous shape of the dragon protein was found to form the shape of a dragon's head by Argonne National Labs as described in their Press Release July 15, 2008.[22]

The mixture of particles in liquid referred to in chemistry as a solution is the ultimate key in making self-assembling crystals. In this mixture, submicron silica spheres float around in the liquid (called the solvent), which is often ethanol. Ethanol causes the silica spheres to crash into a plate that is placed in the solution. As the temperature rises and the ethanol evaporates, the *meniscus* travels down the plate, making it easier for any submicron silica spheres floating nearby to stick to the plate. As time goes by, more and more spheres deposit themselves on the plate, forming an orderly pattern. When all the ethanol is evaporated, the first layer is finished and subsequent layers can be added by repeating the process (thin-film applications). The result: a colloidal self-assembly-colloids being the solution of silica spheres (between 1 nm and 1 micron in size) suspended in a solvent.

By varying the size of the spheres being placed in the solutions, you can introduce useful chemical defects at specified layers. (See Figure 10-10: A) Silicon Nano Spheres in Human Blood with various dye coatings. B) Enlarged Closeup of Silicon Nano Spheres and Nanotubes with specifically Chinese Lantern Nanotubes in Human Blood). Chinese Lantern silicon nanotubes were first developed by Dr. Z. Wang at Georgia Tech.[23, 24]

After all of the spheres are lined up on the plate, they may be coated in a polystyrene plastic – thus allowing one to get it into the dead spaces between spheres. Once that hardens, one may use a chemical etching process to remove the silica spheres, which produces an inverted representation of the original structure as a super lattice.

(See Figure 11-11: Nano Claw, Nano silica spheres, and dragon protein thin layer technology.)

Making these crystals is very quick and relatively easy. It is so easy that many take-home scientists are making them in their homes as DIYbio[25]. Portrayed as techno-progressive, rogue, and, above all, hip, this global cadre of DIYbio practitioners or biohackers are stylized as being capable of doing, at home, what just a few years ago was only possible in the most advanced university, government, or industrial labs. What is clear is that the emergence of DIYbio and synthetic biology add urgency to the creation of a framework for systematically evaluating the risks and dangers of bionanotechnology engineering.[26]

Smart Dust, Smart Crystal Motes and MEMS

Information taken directly from the University of California, San Diego Sailor Institute, adds to the validity of the many applications of smart dust and smart crystal motes. Originally, the term and use of smart dust was developed by its pioneer inventor, Dr. Kris Pister, principal investigator, and co-investigators, Joe Kahn and Bernhard Boser (MLB Co), at the University of California, Berkeley, for autonomous sensing and communication in a cubic millimeter of space as funded by DARPA/MTO MEMS program with possible utilization within the academic network of global pathogen and countermeasures as applied to many of our country's governmental allies.

TARGETED SMART DUST- HOW IT WORKS

In order to spontaneously assemble and orient the micron-sized porous Si "smart dust," we couple chemical modification with the electrochemical machining process used to prepare the nanostructures. The process involves two steps, (see the scheme below). In the first step, a porous photonic structure is produced by etching silicon with an electrochemical machining process. This step imparts a highly reflective and specific color-code to the material that acts like an address, or identifying bar-code, for the particles. The second step involves chemically modifying the porous silicon photonic structure so that it will find and stick to the desired target. In the present case, we use chemistry that will target the interface between a drop of oil in water; but, we hope to be able to apply the methodology to pollution particles, pathogenic bacteria, and cancer cells. The two steps (etch and modify) are repeated with a different color and a different chemistry, yielding two-sided films. The films are broken up into particles about the size of a human hair. With the chemistry shown below, the particles seek out and attach themselves to an oil drop, presenting their red surface to the outside world and their green surface toward the inside of the drop.

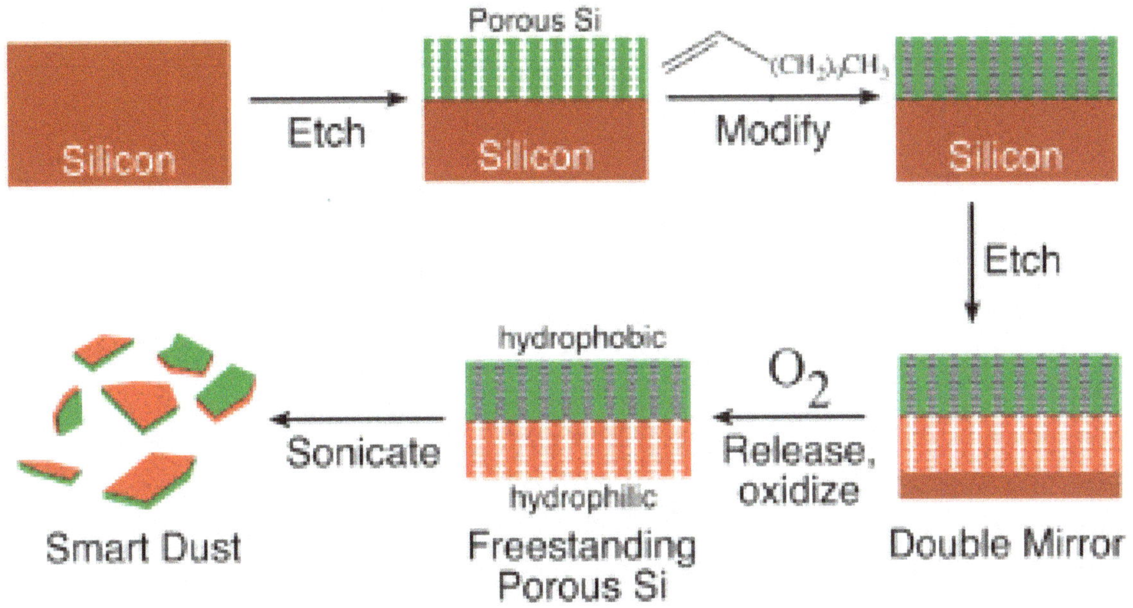

Once they find the interface for which they were programmed, the individual mirrored particles begin to line up, or "tile" themselves on the surface of the target. As an individual, each particle is too small for one to observe the color code. However, when they tile at the interface, the optical properties of the ensemble combine to give a mirror whose characteristic color is easily observed. This collective behavior provides a means of amplifying the molecular recognition event that occurs at the surface of each individual particle.

As a means of signaling their presence at the interface, the particles change color. As the nanostructure comes in contact with the oil drop, some of the liquid from the target is absorbed into it. The liquid only wicks into the regions of the nanostructure that have been modified with the appropriate chemistry. The presence of the liquid in the nanostructure causes a predictable change in the color code, signaling to the outside observer that the correct target has been located. This work was first reported in J. R. Link, and M. J. Sailor, *Proc. Nat Acad. Sci.* 2003 *100*, 10607-10610.[27]

The application of smart nanostructure that caused a change in color – a code – was first observed in the fluorescent dye found in many individuals sclera after exposure to aerial spraying for the brown moth in California in recent years. This was termed the "Eye of Horus Effect," since it patterned the ancient Egyptian mathematic system of addressing the senses of smell, sight, hearing and odor as reported in the 2009 Annual Conference of the National Registry of Environmental Professionals, Journal of Environmental and Sustainability.[28] (See Figure 12-12.)

Smart Dust, and its spin-off technology of smart crystal motes, are miniaturized sensor/transmitters that are sprinkled onto an area - such as a battlefield - and used to analyze the environment. It was originally developed by Professor Kris Pister at the University of California at Berkeley. It is expected, that in the next decade smart dust particles will be be no more than 1 cubic millimeter in size, which includes a solar cell, a sensor, CPU, memory and radio transmitter. This technology has emerged as the mu-chip for Hitachi industries and the Sencil™ technology for Dr. Gerald Loeb at the Alfred E. Mann Institute at UCLA.[29, 30]

Smart dust is being developed as a sensory bioweapon with which one can protect themself from biological and chemical weapons exposure. Yes, the answer may very well be blowing in the wind through the efforts,in 2004, of Dr. Michael Sailor, a professor in the department of chemistry and biochemistry at the University of California, San Diego. He has used the sheen of a beetle's wing as a simple way to explain this technology - the wing does not have pigments, but we see it in different colors due to the various iridescence. The color is produced by two other properties: optical interference – the same phenomenon behind the colors in rainbows and soap bubbles, and in elaborate structures in the wing surfaces of a beetle.

Through DARPA funding, Dr. Sailor intends to make nanoparticles (smart dust) imbedded with iridescent colors into "fingerprints" that can be added to explosives and other chemicals, making it possible to trace a bomb or an illegal drug back to its single manufacturer. He also worked on making these smart nanoparticles – dust – reflect signature colors when they encountered specific pathogens in air or water to create a cheap, disposable sensor for detecting chemical and biological weapons.[31]

Dr. Sailor makes the nanoparticles called smart dust by creating a filter for light in the surface of a silicon wafer about the size of a quarter. He places the wafer in a conductive solution, and then electrochemically corrodes it with an alternating current, which Sailor says, "as (the corrosion) drills down into the silicon, it bottlenecks and opens up again, then bottlenecks and opens up again."[32] The result is a delicate etched network of parallel pores about two nanometers in diameter. Using ultrasound vibrations, Dr. Sailor then crumbles the wafer into particles about the width of a hair.

When the dust is dispersed in air or water, ordinary dust particles scatter light in every direction, but when illuminated with a laser, Dr. Sailor explains, the smart dust appears quite different. "You'll get this one sharp, very precise wavelength of light for a given angle coming in and bouncing off the surface," he says. "The colors that result are incredibly vibrant, strong (and) highly reflective." By varying the current, the length of the process and the composition of the solution, Dr. Sailor can create filters that produce millions of specific colors. Each color is determined by the refractive index of those complex layers in the silicon. He has been further quoted as saying that the refractive index is like a barcode a laser can read to determine the composition of the dust.

Dr. Michael Sailor's work has caught the interest of DARPA because of its battlefield and counter-terror applications as stated in the article by John Harney on March 12, 2003, published in MIT's Technology Review, *Smart Dust Senses Bioweapons*. "The particles could be applied as a 'tag' to certain bomb-making materials, so that when a bomb blows up, investigators can scan a crime scene for the specific smart dust particles." "Most of the stuff that is used in terrorism activities is diverted from legitimate purposes," says Sailor. "If different manufacturers incorporated uniquely coded smart dust, the type of dust found at the bomb scene would indicate where the bomb materials were purchased and provide a clue to the identity of the terrorist who made the bomb."

These same techniques of scientific – forensic – toxicological investigations can be utilized in determining how our environment, global populations and food sources are being exposed to smart dust, G.E.M.S., MEMS, liquid viral crystals, liquid viral envelopes, and other technologies that would only benefit mankind. But in the hands of one of an evil-hearted intent it would cause the demise of humanity as we know it and the rebellion of Mother

Earth in her continual poisoning. Networks utilizing the smart dust can be made with the crystallization of the dust to form a crystal mote through wireless technology, as found in a simple cell phone or telecommunications network system known as Smart DAA's.[33, 34]

Smart Dust, Molecular Forklift, Sensil™ and Lab-on-a-Chip

In January 2009 researchers at the University of Florida, Gainesville, FL developed molecular forklifts, which overcame an obstacle to "smart dust." The researchers observed algae as a livid green giveaway of nutrient pollution as found in a lake. Scientists would love to reproduce the action in tiny particles that would turn algae into different colors if exposed to biological weapons, food spoilage, or signs of poor health in the blood.

The University of Florida engineering researchers have tapped into the working parts of cells to clear a major hurdle to creating site specific "smart dust" for not only bioweapons but cellular disease repair networks. Their new approach to technology, known as "lab-on-a–chip," has been published in the journal *Nature Nanotechnology*.

"Instead of just changing one part of an existing system, we have a new and different way of doing things, "said Henry Hess, UF assistant professor of materials science and engineering and the author of the sensor paper. "And we can do it this way because of building blocks from bionanotechnology, and that's what makes it very exciting."

The researchers coupled the technologies of lab-on-a-chip with forklift systems assembled from natural motor proteins and specific pairs of antibodies to latch onto target contaminants. They do not use electricity to forge the various zones of exposures loaded aboard with fluorescent particles, or tags, but the naturally derived forklifts are powered by adenosine triphsophate, or ATP, the molecule that carries energy to the cells. The key advance, as quoted by Dr. Michael Sailor, co-author of the article, stated that the authors incorporated a transport mechanisms derived from a natural system into an artificial microsensor."

In September 2009 a specimen was isolated from the platform scaffolding technology of nasal sensory technology that was utilized in the coupling of Sencil™ technology, smart dust/crystals and the utilization of mass population in selected cities to monitor their environment (remotely). See Figure 13-13, which shows the field isolation of Sencil™ technology as developed by the Alfred E. Mann Institute, UCLA for sensory monitoring applications. The specimen was isolated from the nasal bulb of a female who was exposed to aerial spraying during September 2009 in Los Angeles, CA. It was also documented by the individual that continual EMF or RDIF frequencies were being emitted at specific times and frequencies, thus confirming the application of remote motes stimulating the growth of advanced nanomaterials into specific nanomachines, nanorobots, or other similar tools/devices.

In the December 20, 2009, issue of *Nature Nanotechnology,* it was stated that scientists at the U.S. Department of Energy's Brookhaven National Laboratory have found a new way to use a synthetic form of DNA to control the assembly of nanoparticles – this time resulting in switchable, three-dimensional, and small-cluster structures that might be useful, for example, as biosensors, in solar cells and as new materials for data storage.

The Brookhaven team, lead by physicist Oleg Gang, has been refining techniques to use strands of artificial DNA as a highly specific kind of Velcro, or glue, to link up nanoparticles, thus creating "smart glue." These same techniques have been applied to the new Harvard-designed DNA folding protein origami technologies to compress multiple materials into a single "smart dust platform."

This type of technology was isolated from an individual who was shedding black specs with a clear plastic margin around the specimen. The individual had been exposed to multiple vaccines and had exposure to bed bugs (scabies). (This may have been synthetic biological delivery systems under pathogen countermeasures population testing, since many individuals experienced this after staying in a hotel.) Further investigations may determine that the advanced nanomaterials may have been part of the detergents used to launder linen in these facilities. See Figure 14-14, which is a photomicrograph after a static charge from the dissecting needle used to align the specimen for photographing was enough to cause the DNA origami folding protein to unfold. Further analysis utilizing Micro FTIR technologies showed four specific analyses of high density polyethylene, acrylonitriles, cotton, and polyesters. Further analysis of the dye showed it to be composed of a dye that has 56 letters and manufactured in Russia.

Summation and Conclusions

Nanotechnology investors, military, academia, and industry are the partners of the global rise in the many different benefits for the utilization of nanotechnology. Through this rise of phenomenal technology under the auspices of benefiting mankind and the environment, one can only pray to a higher being that through its release into the environment, planned or accidental, remedies will be made for its antidote upon exposure. The use of silane, alone, as a tracking compound with smart dust and/or antimicrobial coating on nanoparticles, can be antidoted in the human body with the use of Opaline Dry Oxy Granules and Dr. Willard's Catatlytic Altered Water. [35] It will decompose and the silicon/silica by-products will be released from the lymphnodes. Silicon nanotubes react with variable pH's, which can be further mediated with the application of alkalinized water as found in Kangen/Enigic alkalinizing water products.

Looking again at Figure 10-10, one can only wonder who is the true mastermind - a mastermind who created something that can be nearly invisible to the naked eye, yet become our greatest national security threat, as it can cause the death of mankind with a simple flip of a switch, as nanotechnology advances into the realms of synthetic biology and nanovectored gene delivery systems. Time will only tell us after She has run through the universe with Eternity, holding their hands together, as they are looking over their shoulders at Mother Earth and her children, saying, "Where Have They All Gone?"

Hildegarde Staninger, Ph.D., RIET-1

Industrial Toxicologist/IH & Doctor of Integrative Medicine

REFERENCES

1. Williams, Linda and Dr. Wade Adams. Nanotechnology Demystified: A Self-Teaching Guide. The McGraw-Hill Companies. New York. © 2007

2. Marshall, Peter. The Magic Circle of Rudolf II: Alchemy and Astrology in Renaissance Prague. Walker & Company. New York. © 2006.

3. www.nanotechtraining.com (Nanotechnology in India - Leaders in Nanotechnology Learning and Knowledge Sharing) © 12/23/2009

4. Wei, Chiming. Nanomedicine. Medical Clinics of North America. Volume 91 Number 5. the clinics.com Elsevier Saunders. Philadelphia, PA. © Sept. 2007

5. Williams, Linda and Dr. Wade Adams. Nanotechnology Demystified: A Self-Teaching Guide. The McGraw-Hill Companies. New York © 2007 pgs 143-147.

6. Press Release © 2007 Massachusetts Institute of Technology - Soldier Institute. Nano Worm developed at MIT. Staninger, Hildegarde. Project: Fiber, Meteorite and Morgellons. Phase 1-4.
www.rense.com/morgphase/phase2_1.htm © 2006 (A privately funded project.)

7. Staninger, Hildegarde. Toxicological Dysfunction Analysis™ (TDA): Mr. James Walbert. Integrative Health Systems, LLC, Los Angeles, CA © December 24, 2009

8. Booker, Richard and Earl Boysen. Nanotechnology for Dummies: A Reference for the Rest of Us! Wiley Publishing, Inc. Hoboken, NJ. © 2005

9. Booker, Richard and Earl Boysen. Nanotechnolgy for Dummies: A Reference for the Rest of Us! Wiley Publishing, Inc. Hoboken, NJ. © 2005 pgs 168-170.

10. Soghoian, Marshall. Page One Science. Alexandria, Virginia © 2001 meeting Anasell Group, Inc. for Sensory Technologies and Issues of National Security.

11. Link, Jamie R. and Michael J. Sailor. Smart dust: Self-assembling, self-orienting photonic crystals of porous Si. National Academy of Science. USA 100(19): 10617-10610. Washington, D.C. © September 16, 2003.

12. www.ucla.com Alfred E. Mann Institute. Sencil™ developed by Dr. Gerald Loeb. Technical papers and technology for diabetes monitoring.

13. Staninger, Hildegarde. Toxicological Dysfunction Analysis™ Report. Dr. Faranak Behi (and Massachusettes General Casulalty Insurance, Boston, MA) © 2007. DNA Tags- Black Panther/Pansey; Dragon Protein, Piezo Electrical Device and Acrylonitrile Nano Thin Film Technology. Integrative Health Systems, LLC, Los Angeles, CA © 2007, 2008, and 2009

14. Staninger, Hildegarde. IBID © 2008

15. Staninger, Hildegarde. IBID © 2009

16. Whipple, Charles. Korean Researchers Achieve Electrically-Pumped Photonic Crystal Single - Cell Laser. From OEMagazine. Feb. 2005 SPIE News Room: DOI: 10.1117/2.5200502.0002

17. Williams, Linda and Dr. Wade Adams. Nanotechnology Demystified: A Self Teaching Guide. McGraw Hill Publishing, Inc. Hoboken, NJ © 2005 Chapter 8: Materials, pages 144 - 149.

18. Williams, Linda and Dr. Wade Adams. IBID © 2005

19. Staninger, Hildegarde. "*Nanocomposite Material Exposure through Aerial Emissions cause Acetylcholinesterase Inhibition*." National Registry of Environmental Professionals (NREP). 2009 Annual Conference. Journal of Environment and Sustainability. October 2009.

20. www.sciencedaily.com/release/2004/11///041108012459.htm USCD Chemist Use Tiny "chaerones: to direct Molecules and Nanoparticles in Drop of Liquid.

21. Xie, Yougshu; Hill, Johnathan P.; Chavet, Richard, Ariga, Katshiko. "*Porphyrin Colorimetric Indicator in Molecular and Nano-Architectures.*" Journal of Nanoscience and Nanotechnology. Vol. 7, Number 9, September 2007 pg: 2969-2993(25). American Scientific Publishers.

22. www.argonnenationallabs.com DOE/Argonne National Labs Press Release 15-July- 2008. Newly Described "Dragon Protein" could be key to Bird Flu Cure.

23. www.nanoscience.gatech.edu/zlwang/publication.html Silicon Nanotubes, Nanofibers and Micoarrays.

24. Wang, Zhong, Gao, Rui Ping, Pan, Zheng Wei, and Zu R. Dai. "*Nano Scale Mechanics of Nanotubes, Nanowires and Nanobelts*." Advanced Engineering Materials. © 2001, 3 No. 9.

25. Bennett, Gaymon; Gilman, Nils; Stavrianakis, Anthony, and Paul Rabinow. "*From synthetic biology to biohacking: are we prepared*?" Nature Biotechnology. Volume 22 Number 12 December 2009 pgs: 1109-1111.

26. de S. Cameron, Nigel M. and Arthur Caplan. "*Our Synthetic Future – a Commentary: Two prominent ethicists provide their views on the ethical debates surrounding synthetic biology*." Nature Biotechnology Volume 27 No. 12 December 2009 pages 1103-1105.

27. www.ucsd/sailorinstitute.com smart dust – © 2009

28. Staninger, Hildegarde. "*The Eye of Horus Effect.*" National Registry of Environmental Professionals (NREP). 2009 Annual Conference. Journal of Environment and Sustainability © October 2009.

29. www.hitachi.com/new/cnews/////030902.htal Hitachi Develops New RFID with Embedded Antenna u-Chip. Press Release September 2, 2003.

30. http://ami.usc.edu/projects/ami/projects/sencil/ Video Overview, Technology Overview, Glucose Sensor Development, Cancer Applications and other Publications © 2009

31. Harney, John. "*Smart Dust Senses Bioweapons.*" http://www.technologyreview.comcomputing13121/ Technology Review © March 12, 2003.

32. Lieberman, Bruce. "*The rapidly advancing science is forecast to transform society.*" (Spotlight on Nanotechnology) The San Diego Union-Tribune. © March 12, 2005.

33. http://pda.physorg.com_news 180624054.html Switchable Nanostructures Made with DNA © December 21, 2009, Nanotechnology/Bio & Medicine.

34. www.dna.caltech.edu/~pwkr/
DNA origami folding protein

35. www.wikipedia.com "Silane" "Silane breaks down in the presence of oxygen, forming carbon dioxide and silica." © 2009 Soft clinical trials conducted at Integrative Health Systems for Opaline Solutions, LLC, Tucson, AZ, manufacture of a product known as Opaline Dry Oxy Granules degradated nano composite materials coated with silane as it made Oxygen (O_2) and water. © 2007 H. Staninger. Dr. Willard's Catalytic Altered Water (CAW) is manufactured by CAW Industries, Inc., Rapid City, S.D. and has an inert Natural affinity for silica. It breaks it down larger molecules through molecular bonding reactions as associated in reverse micelle reactions of covalent bonded materials. © 1986, Hy-Tox, Inc. www.krystheraputicwater.com Enagic/Kangen Alkaline water systems. Individuals utilizing Dr. Hildegarde Staninger's Cellular Detoxification Program had a 20 to 40% increase of cellular detoxification and body fat loss due to drinking Kangen water on a daily routine for 90 days.

Figure 1-1: Diagram of Self Assembled Carbon Nanotubes.

Figure 2-2: Actual Field Specimen form a Female Exposed to Advanced Nano Material (diagnosed Morgellons). Project: Fiber, Morgellons & Meteorite ©2006 H. Staninger, *Photomicrograph taken by Dr. Rahim Karjoo, Pathologist*. Actual Field Specimen of Nanotubes, Nanoclaws, and Sol-Gel (hydrogels).

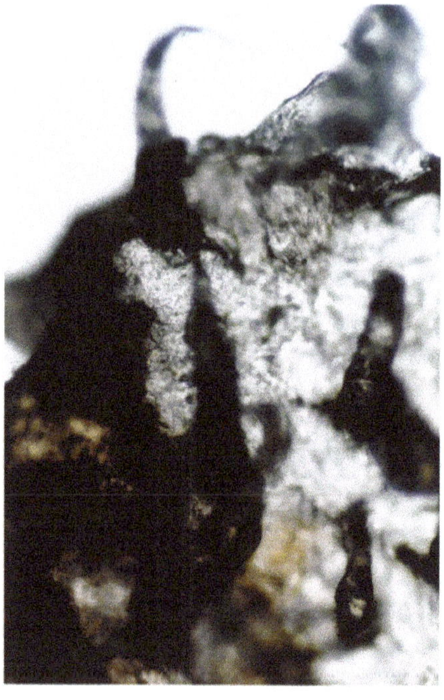

Figure 3-3: "Goldenhead" Taken from Project: Fiber, Morgellons & Meteorite © 2006 H.Staninger, *Photomicrograph taken by Dr. Rahim Karjoo, Pathologist*

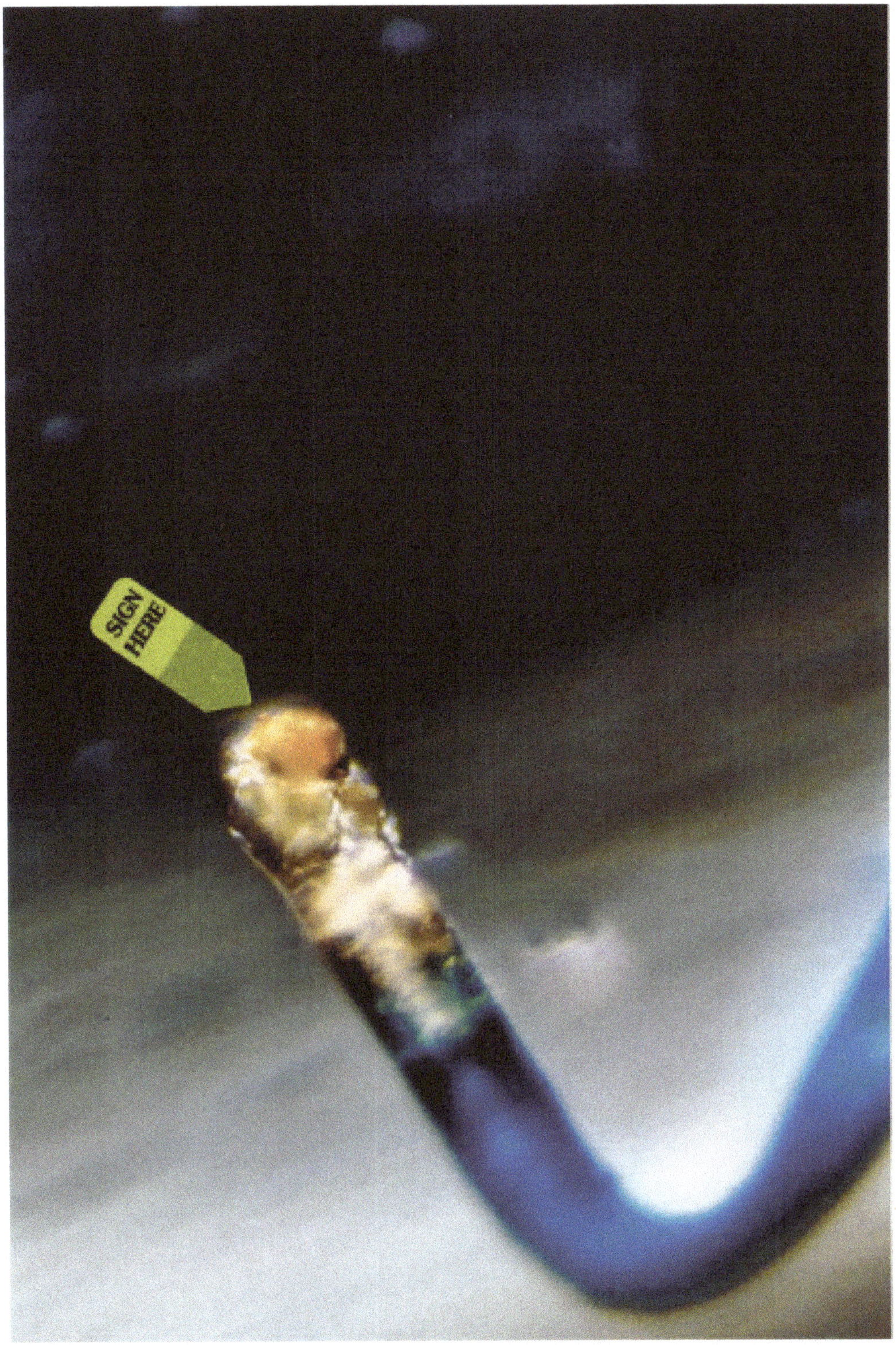

Yellow pointer notes sensor and/or viral digital camera when coupled with high density polyethylene nanotube lens.

Figure 4-4: Image of Self-Developing Nano Composite Material After 21 days extracted from Female Exposed to Advanced Nano Materials. (Note: Nano Claws developed by Georgia Institute of Technology, Dr. Z. Wang.) *Photomicrograph taken by Dr. Rahim Karjoo*, Project: Fiber, Morgellons and Meteroite © H. Staninger)

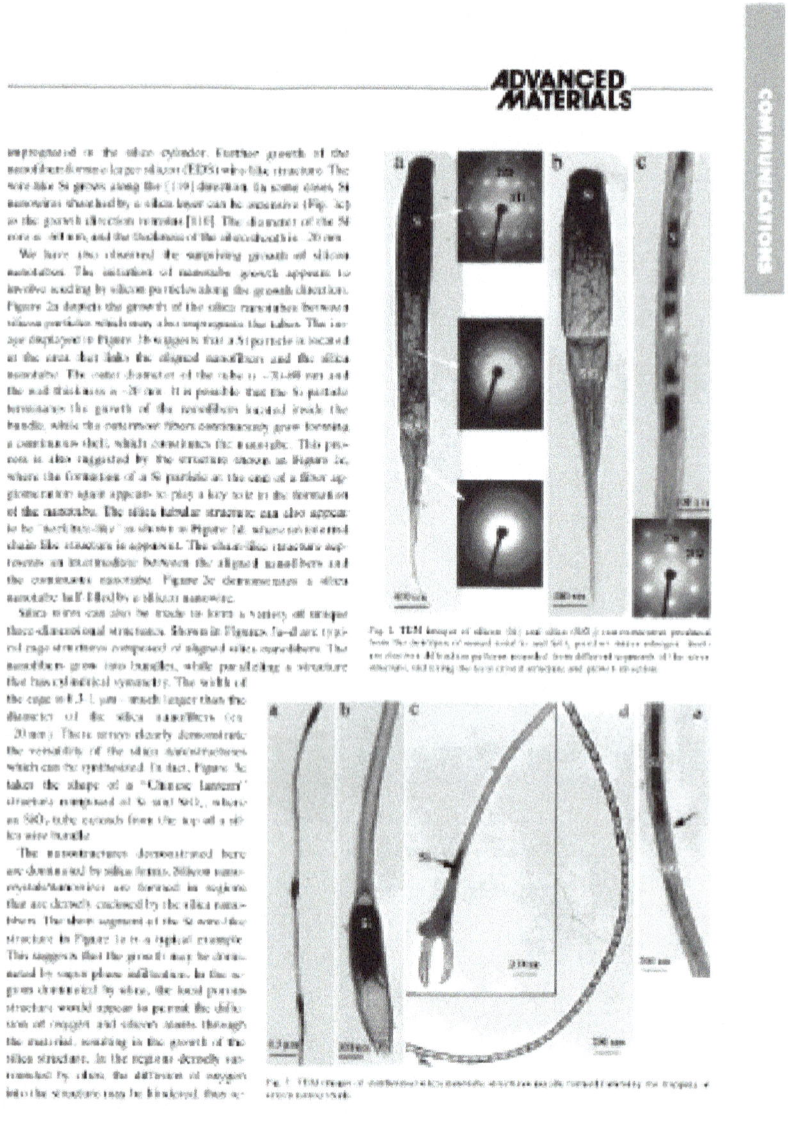

ADVANCED MATERIALS

COMMUNICATIONS

Figure 5-5: Nano-Light Emitting Diode from Sencil™ Technology as found in the Field from a Female After Exposure to Aerial Spraying, Los Angeles, September 2009.

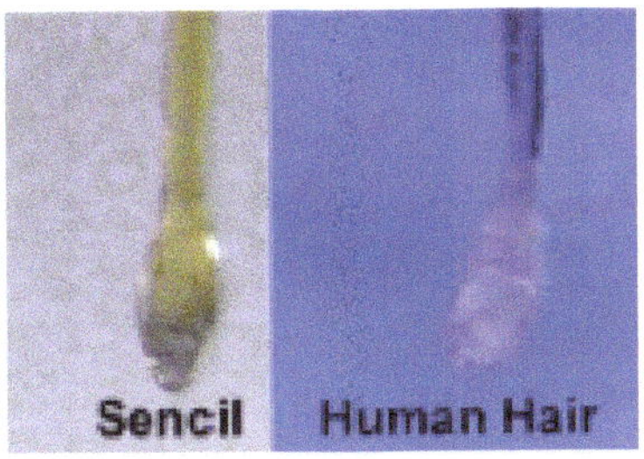

Sencil™ Image take from Alfred E. Mann Institute's website. University of Southern California

http://ami.usc.edu/projects/ami/projects/sencil

Sencil™ technology removed from nasal bulb of female after aerial spraying in Los Angeles, CA September 2009. Note tip is "red" in color, where sensor is located.

Figure 6-6: Silane Based Nano Composite Material (2 polymer materials and silicone head).

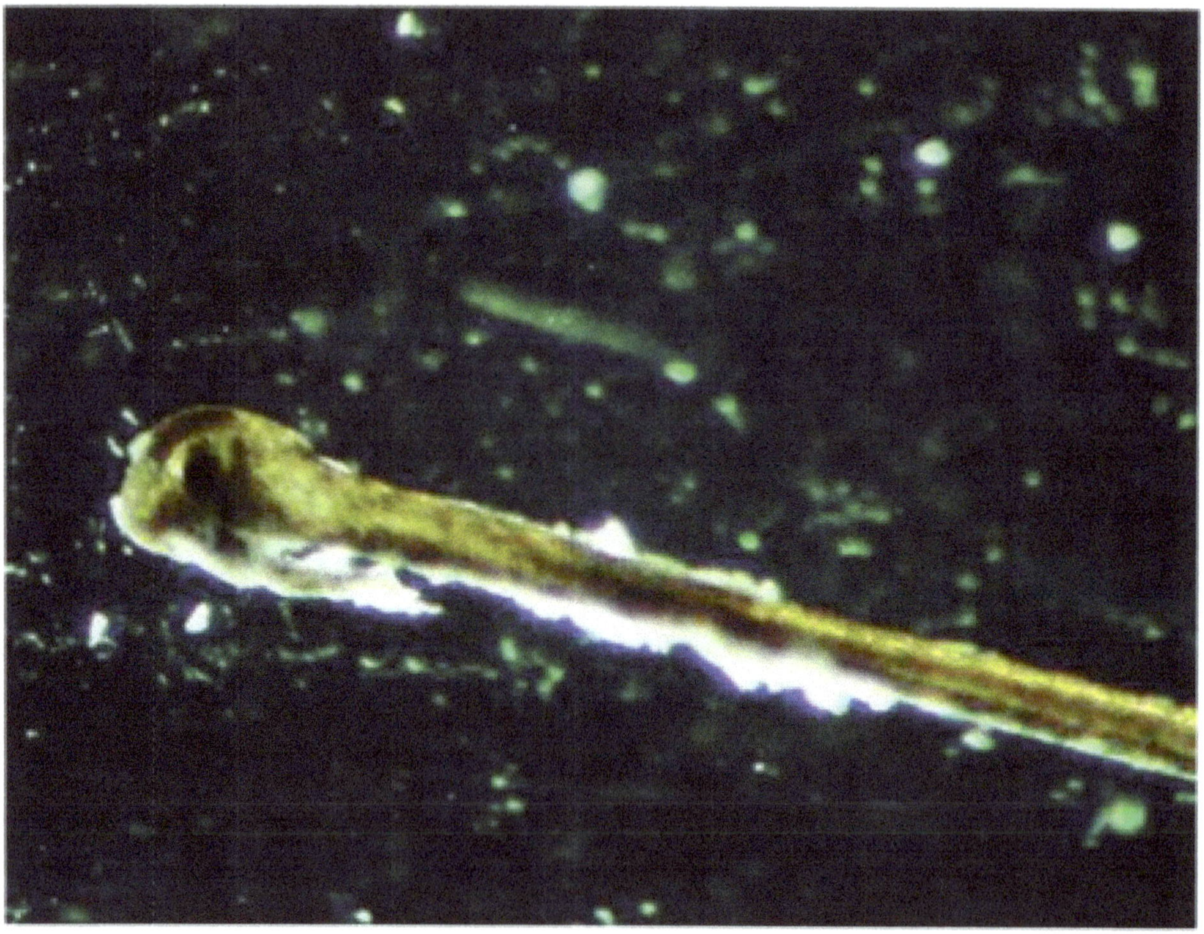

Figure 7-7: Acrylic Nano Composite Material with Fibers of a Piezo Electrical Device. Specimen originally located in Female's Throat after she coughed it up. (Note: Fiber Twists and metal composites as spheres placed 2 per space.) *Photomicrograph taken by Applied Consumer Services, Inc. © 2007 H. Staninger*

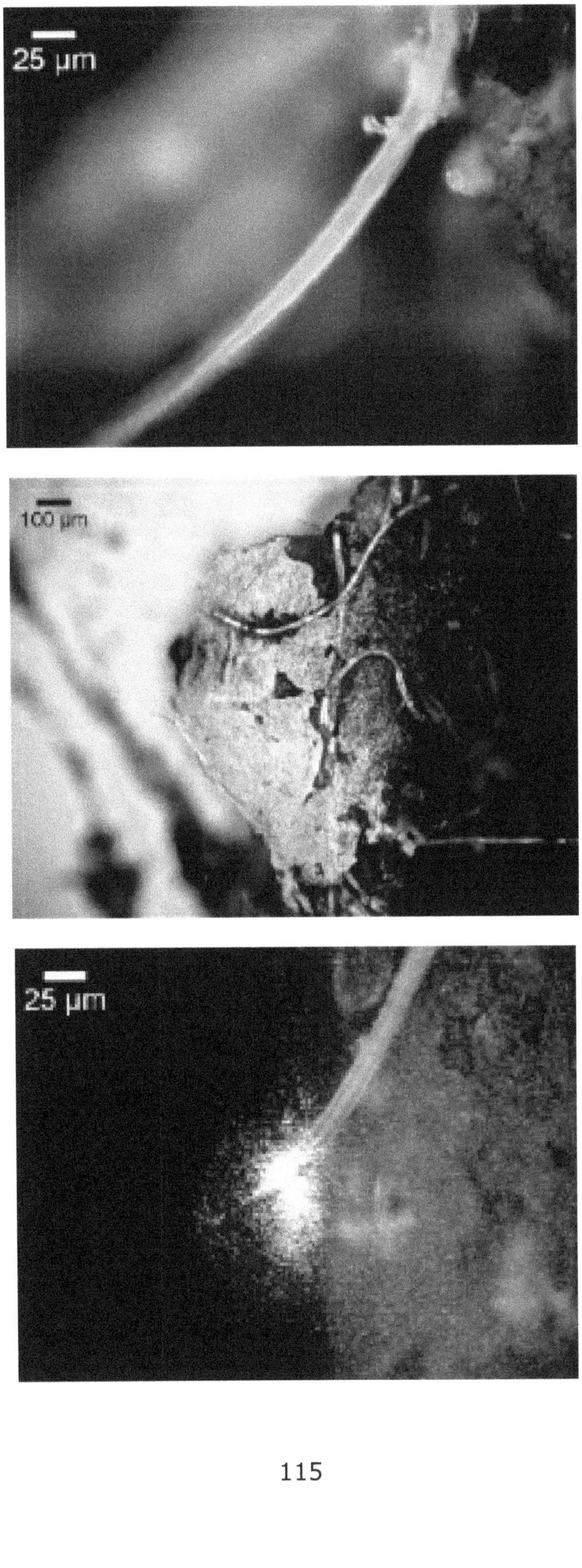

Figure 8-8: Edible Braille RDIF Chip as Compared to Granules of Sugar.

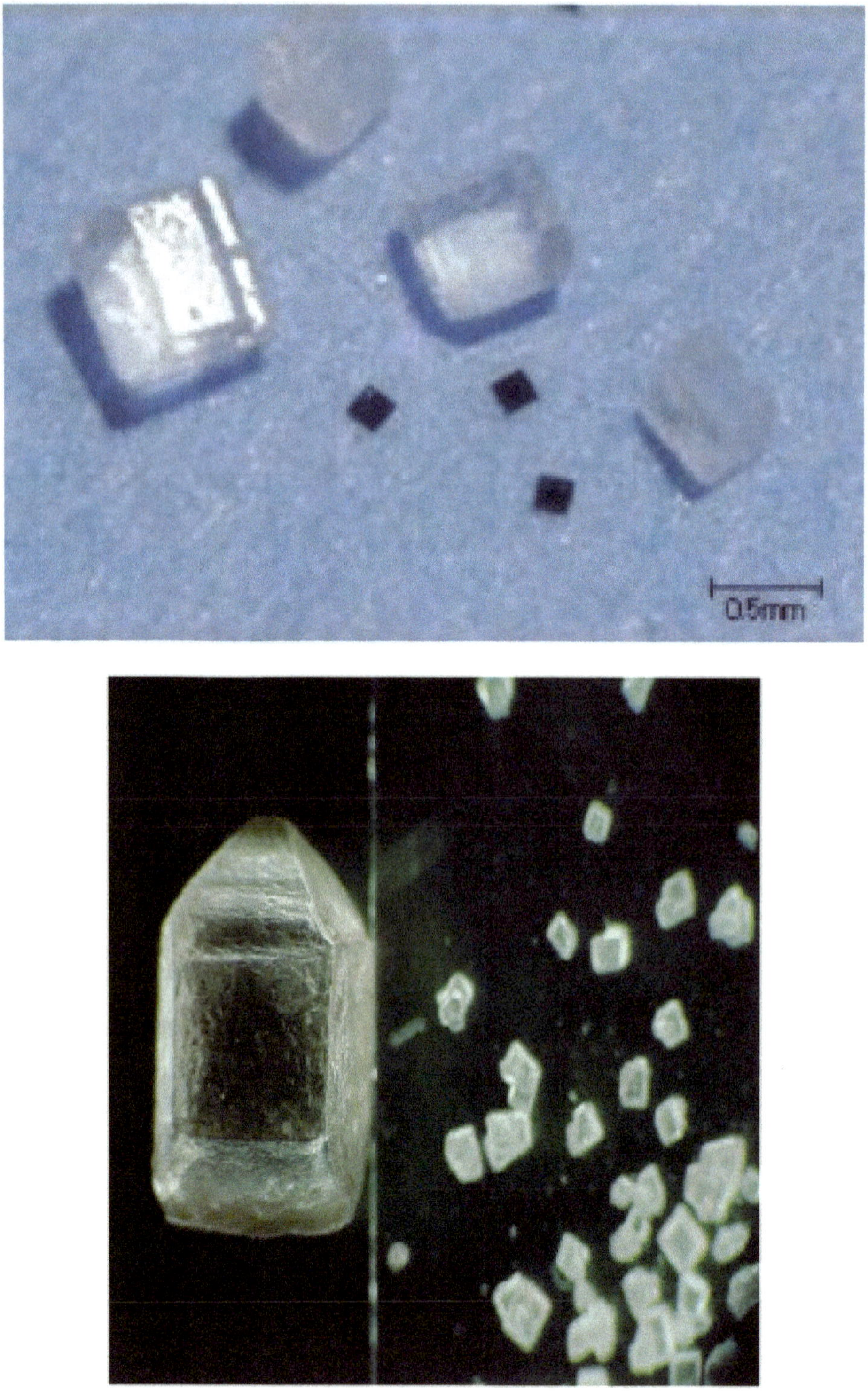

Note: Edible Braille Chip to left as compared to sugar granules to right in above photo micrograph. Photomicrograph by Applied Consumer Services, Inc. © 2009 H. Staninger

Figure 9-9: Note waterborne polyurethane nano composite material forming styrene nanohorns on human skin surface. *Photomicrograph taken by Dr. Rahim Karjoo,* PathologistProject: Fiber, Morgellons and Meteorite © 2006 H. Staninger
(Note: Nanohorn non pinkish area, upper left corner of photomicrograph.)

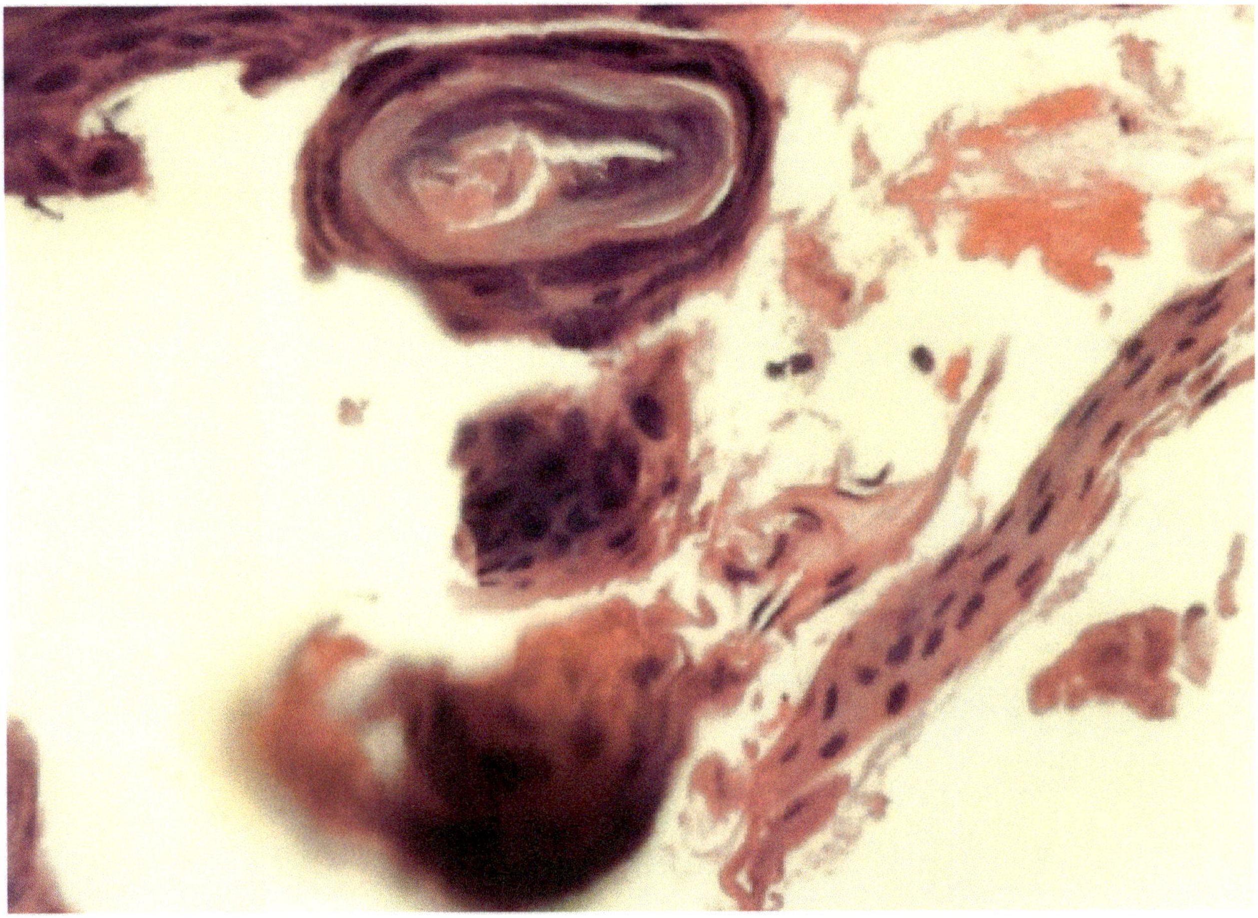

Figure 10-10: A) Silicon Nanospheres in Human Blood with various dye coatings with Chinese Lantern Nanotubes in Human Blood. Taken from Project: Fiber, Morgellons & Meteorite. © 2006 H. Staninger. *Photomicrograph taken by Dr. Rahim Karjoo, Pathologist.*

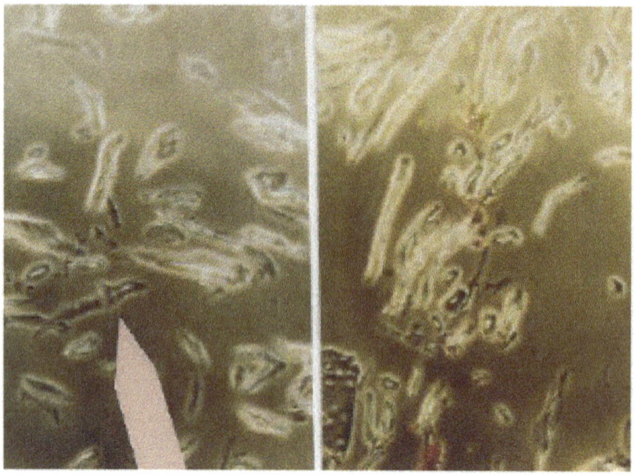

Figure 10-10: B) Chinese Lantern Nanotubes as diagramed by Dr. Z. Wang in his article Silica Nanotubes and Microarrays, Advanced Materials Engineering.

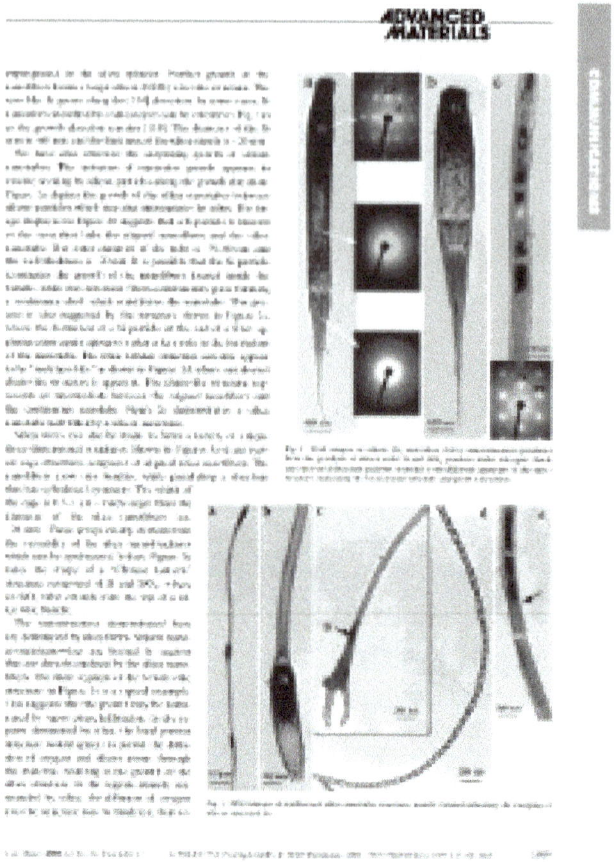

ADVANCED MATERIALS

Figure 11-11: "Dragon Puppy" showing Dragon Protein, Nano silica spheres, and Thin Layer Nano Films/Technology

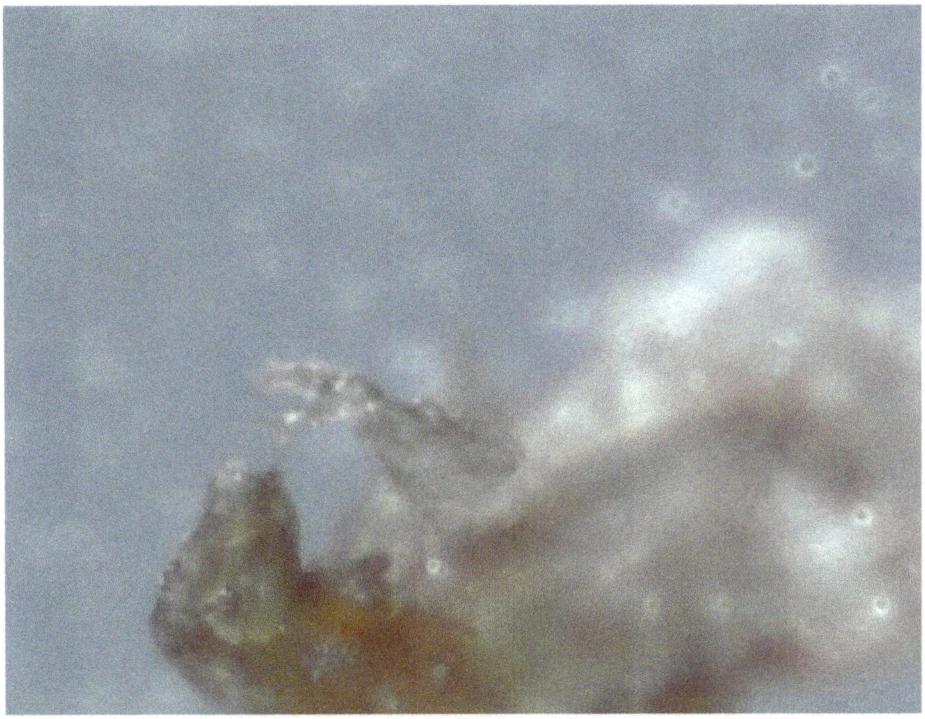

Below Close up of nano claw over front of head. Note silica nano micro arrays as a grid.

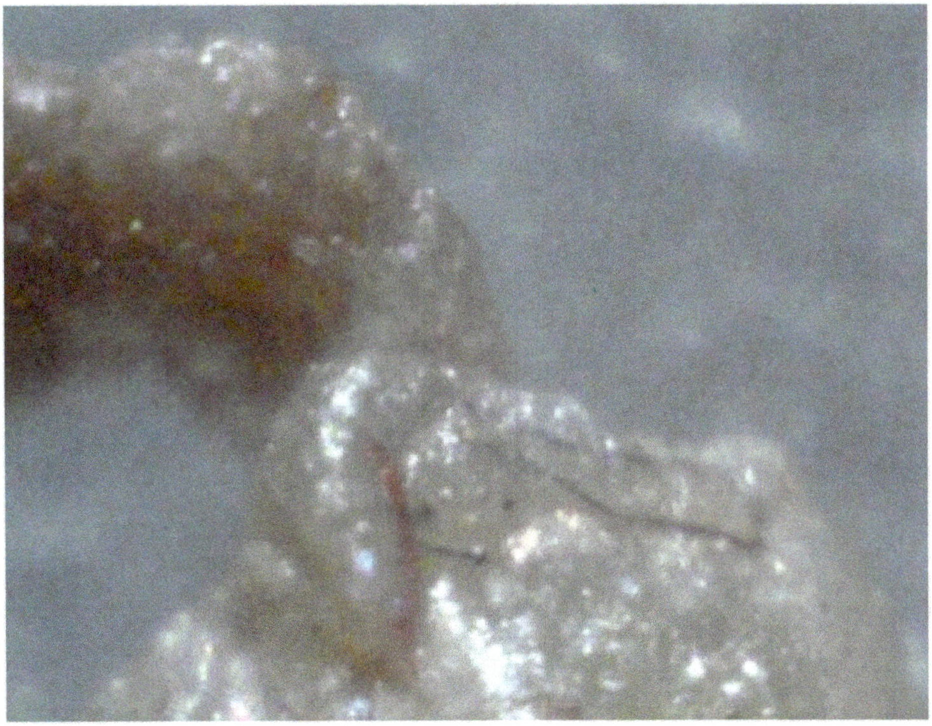

Figure 12-12: Note "Eye of Horus Effect" as published in the 2009 National Registry of Environmental Professional Annual Conference, Journal of Environment and Sustainability. "Eye of Horus Effect" by Hildegarde Staninger, Ph.D., RIET-1, © October 2009

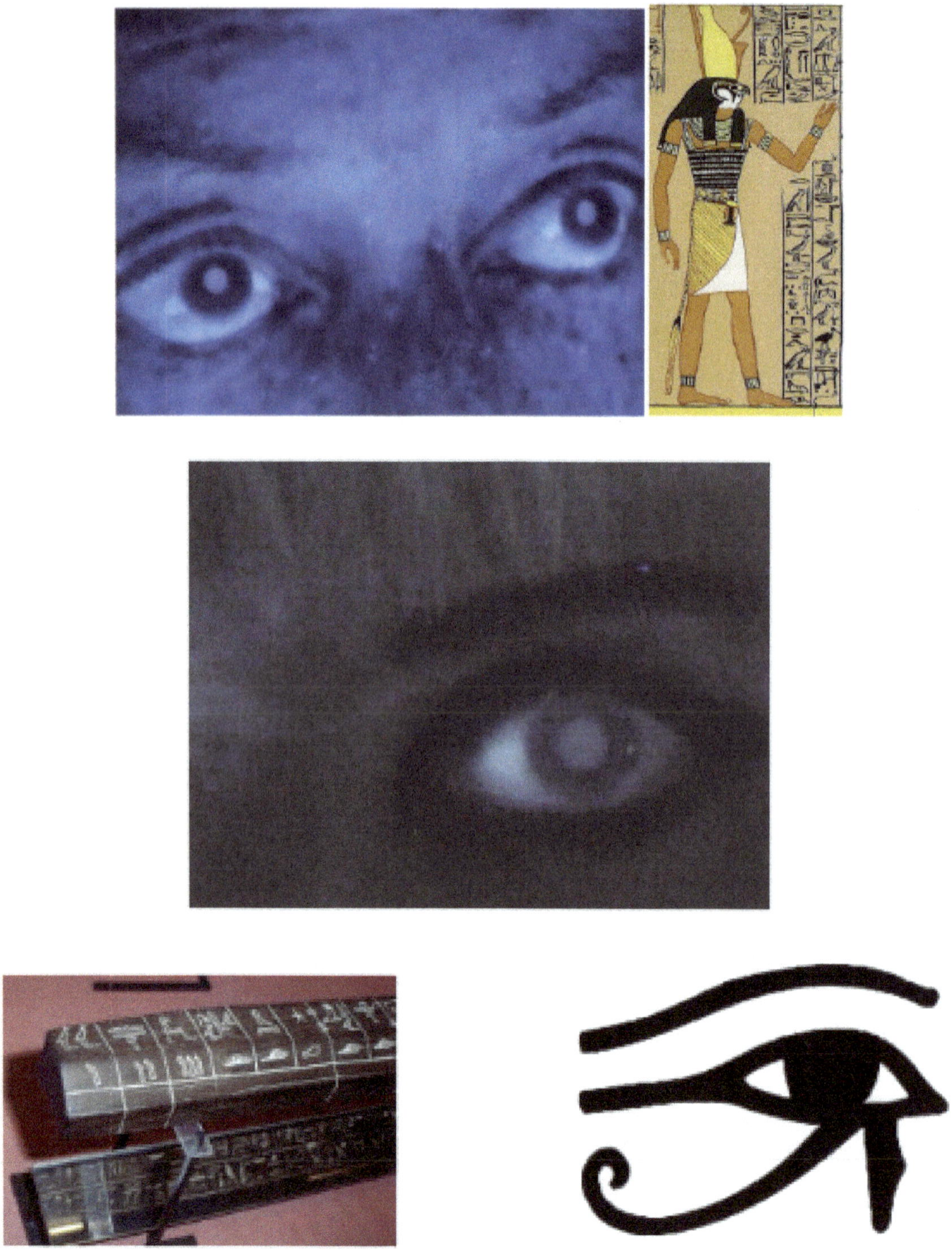

Figure 13-13: Integration of Nasal Sensory Technology, Sencil™, Smart Dust and Crystal Motes. Compare with Figure 5-5.

Figure 14-14: Field Specimen sent to Applied Consumer Service, Inc. laboratory upon Chemist using a dissecting needle to align specimen on slide for photomicrograph, the static charge from the needle caused the original black spec to unfold into 4 distinct fibers (Red, White (2) and Blue). Raman Analysis revealed: high density polyethylene, dacron, acrylin, and cotton with a red dye (56 letters long) called 2-(5-Bromo-2 pyridylazo)-5[N-n-propyl-N-(3-sulfopropyl)amino]aniline, sodium salt, which is currently manufactured in Russia.

Note: nano anchors, appear as "barbed wire."

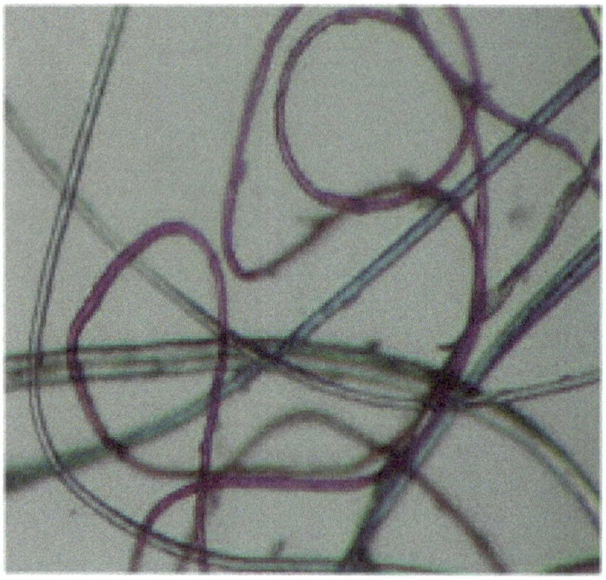

Unfolded DNA Proteins

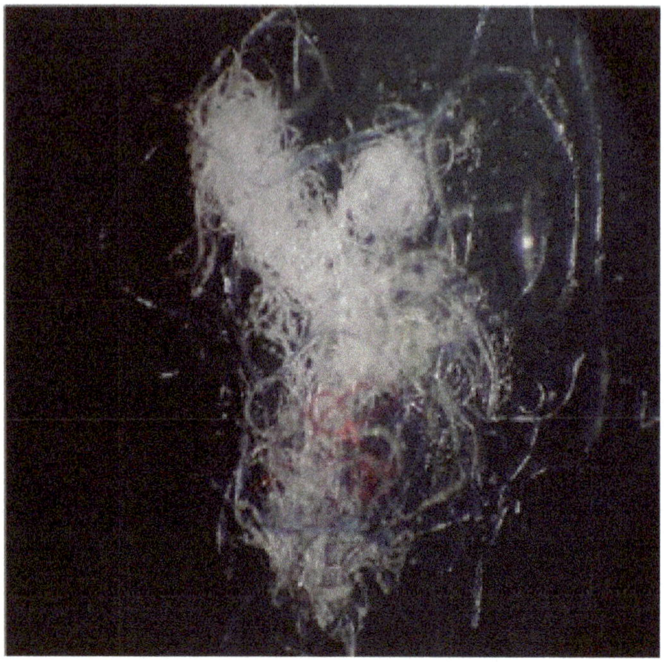

INDEX

A

B

N

O

P

T

www.ingramcontent.com/pod-product-compliance
Ingram Content Group UK Ltd.
Pitfield, Milton Keynes, MK11 3LW, UK
UKHW050408220525
6015UKWH00035B/289

9 781498 492003